HOW TO
GET IDEAS

HOW TO GET IDEAS

Second Edition

Jack Foster

Illustrations by Larry Corby

BERRETT-KOEHLER PUBLISHERS, INC.
San Francisco
a BK Life book

Berrett-Koehler Publishers, Inc.
235 Montgomery Street, Suite 650
San Francisco, CA 94104-2916
Tel: (415) 288-0260   Fax: (415) 362-2512   www.bkconnection.com

Ordering Information
Quantity sales. Special discounts are available on quantity purchases
by corporations, associations, and others. For details, contact the
"Special Sales Department" at the Berrett-Koehler address above.

Individual sales. Berrett-Koehler publications are available through most
bookstores. They can also be ordered directly from Berrett-Koehler.
Tel: (800) 929-2929; Fax: (802) 864-7626; www.bkconnection.com

Orders for college textbook/course adoption use. Please contact
Berrett-Koehler. Tel: (800) 929-2929; Fax: (802) 864-7626.

Orders by U.S. trade bookstores and wholesalers. Please contact Ingram
Publisher Services, Tel: (800) 509-4887; Fax: (800) 838-1149; E-mail:
customer.service@ingrampublisherservices.com; or visit www
.ingrampublisherservices.com/Ordering for details about electronic
ordering.

Berrett-Koehler and the BK logo are registered trademarks of Berrett-
Koehler Publishers, Inc.

Printed in the United States of America

Berrett-Koehler books are printed on long-lasting acid-free paper.
When it is available, we choose paper that has been manufactured by
environmentally responsible processes. These may include using trees
grown in sustainable forests, incorporating recycled paper, minimizing
chlorine in bleaching, or recycling the energy produced at the paper mill.

Library of Congress Cataloging-in-Publication Data
Foster, Jack, 1930–
    How to get ideas / by Jack Foster ; illustrated by Larry Corby. — 2nd ed.
        p.   cm.
    Includes bibliographical references and index.
    ISBN: 978-1-57675-430-6 (pbk.)
    1. Authorship.   2. Advertising.   3. Business writing.   I. Title.
PN147.F66 2006
808'.066659—dc22   2006033541

Second Edition
12 11 10 09   10 9 8 7 6 5 4 3

Text design by Detta Penna

To the three best ideas
I ever had—
My wife, Nancy,
and my sons,
Mark and Tim

Contents

Preface   ix
Acknowledgments   xiii

Introduction: What Is an Idea?   1

Part I: Ten Ways to Idea-Condition Your Mind   12
   1. Have Fun   15
   2. Be More Like a Child   25
   3. Become Idea-Prone   35
   4. Visualize Success   51
   5. Rejoice in Failure   59
   6. Get More Inputs   67
   7. Screw Up Your Courage   83
   8. Team Up with Energy   93
   9. Rethink Your Thinking   101
  10. Learn How to Combine   117

Part II: A Five-Step Method for Producing Ideas   129
  11. Define the Problem   131
  12. Gather the Information   145
  13. Search for the Idea   157
  14. Forget about It   165
  15. Put the Idea into Action   173

Notes   185
Index   199
About the Author   211
About the Illustrator   213

Preface

For seven years I helped teach a 16-week class on advertising at the University of Southern California. The class was sponsored by the AAAA—American Association of Advertising Agencies—and was designed to give young people in advertising agencies an overview of the profession they had chosen.

One teacher talked about account management. One teacher talked about media and research. And I talked about creating advertising.

I talked about ads and commercials, about direct mail and outdoor advertising, about what makes good headlines and convincing body copy, about the use of music and jingles and product demonstrations and testimonials, about benefits and type selection and target audiences and copy points and subheads and strategy and teasers and coupons and free-standing inserts and psychographics and on and on and on.

And at the end of the first year I asked the graduates what I should have talked about but didn't.

"Ideas," they said. "You told us that every ad and every commercial should start with an idea," one of them wrote, "but you never told us what an idea was or how to get one."

Well.

So for the next six years I tried to talk about ideas and how to get them.

Not just advertising ideas. Ideas of all kinds.

After all, only a few of the people I taught were charged with coming up with ideas for ads and commercials; most were account executives and media planners and researchers, not writers and art directors. But all of them—just like you and everybody else in business and in government, in school and at home, be they beginners or veterans—need to know how to get ideas.

Why?

First, new ideas are the wheels of progress. Without them, stagnation reigns.

Whether you're a designer dreaming of another world, an engineer working on a new kind of structure, an executive charged with developing a fresh business concept, an advertiser seeking a breakthrough way to sell your product, a fifth-grade teacher trying to plan a memorable school assembly program, or a volunteer looking for a new way to sell the same old raffle tickets, your ability to generate good ideas is critical to your success.

Second, computer systems are doing much of the mundane work you used to do, thereby (in theory at least) freeing you up—and indeed, requiring you—to do the creative work those systems can't do.

Third, we live in an age so awash with information that at times we feel drowned in it, an age that demands a constant stream of new ideas if it is to reach its potential and realize its destiny.

That's because information's real value—aside from helping you understand things better—comes only when it is combined with other information to form new ideas: ideas that solve problems, ideas that help people, ideas that save and fix and create things, ideas that make things better and cheaper and more useful, ideas that enlighten and invigorate and inspire and enrich and embolden.

If you don't use this fortune of information to create such ideas, you waste it.

In short, there's never been a time in all of history when ideas were so needed or so valuable.

The first edition of this book contains most of what I told my students about ideas.

This second edition:

• Contains two new chapters—5, Rejoice in Failure, and 8, Team Up with Energy—that were suggested by friends and by teachers and students who used the first edition as a textbook.

• Updates some of the examples and references and quotations to make the book more current.

• Is reorganized to make more clear the two parts of the book—*Part I: Ten Ways to Idea-Condition Your Mind,* and *Part II: A Five-Step Method for Producing Ideas.*

Acknowledgments

I learned something about ideas from just about everybody I ever taught or worked with. Any attempt to remember and name them all would fail. A sincere but sweeping "Thank you, everyone" must therefore suffice.

Special thanks go to Tom Pflimlin, whose many suggestions helped me improve the first edition of this work; to Henry Caroselli and Mel Sant, whose many suggestions helped me improve this second edition; to Steven Piersanti and his staff, whose enthusiasm and knowledge and skill helped me transform a rough manuscript into a finished book, and a successful first edition into an even better second edition; and to my family, whose faith sustains me.

Introduction

What Is an Idea?

I know the answer. The answer lies within the heart of all mankind! What, the answer is twelve? I think I'm in the wrong building.

Charles Schultz

I was gratified to be able to answer promptly, and I did. I said I didn't know.

Mark Twain

If love is the answer, could you please rephrase the question?

Lily Tomlin

Before we figure out how to get ideas we must discuss what ideas are, for if we don't know what things are it's difficult to figure out how to get more of them.

The only trouble is: How do you define an idea? A. E. Housman said: "I could no more define poetry than a terrier can define a rat, but both of us recognize the object by the symptoms which it produces in us." Beauty is like that too. So are things like quality and love.

And so, of course, is an idea. When we're in the presence of one we know it, we feel it; something inside us recognizes it. But just try to define one.

Look in dictionaries and you'll find everything from: "That which exists in the mind, potentially or actually, as a product of mental activity, such as a thought or knowledge," to "The highest category: the complete and final product of reason," to "A transcendent entity that is a real pattern of which existing things are imperfect representations."

A lot of good that does you.

The difficulty is stated perfectly by Marvin Minsky in *The Society of Mind*:

> Only in logic and mathematics do definitions ever capture concepts perfectly. . . . You can know what a tiger is without defining it. You may define a tiger, yet know scarcely anything about it.

If you ask people for a definition, however, you get better answers, answers that come pretty close to capturing both the concept and the thing itself.

Here are some answers I got from my coworkers and from my students at the University of Southern California and the University of California at Los Angeles:

> It's something that's so obvious that after someone tells you about it you wonder why you didn't think of it yourself.

> An idea encompasses all aspects of a situation and makes it simple. It ties up all the loose ends into one neat knot. That knot is called an idea.

> It is an immediately understood representation of something universally known or accepted, but conveyed in a novel, unique, or unexpected way.

> Something new that can't be seen from what preceded it.

It's that flash of insight that lets you see things
in a new light, that unites two seemingly
disparate thoughts into one new concept.

An idea synthesizes the complex into the
startlingly simple.

It seems to me that these definitions (actually,
they're more descriptions than definitions, but no
matter—they get to the essence of it) give you a better
feel for this elusive thing called an idea, for they
talk about synthesis and problems and insights and
obviousness.

The one that I like the best, though, and the one
that is the basis of this book, is this one from James
Webb Young:

An idea is nothing more nor less
than a new combination of old elements.

There are two reasons I like it so much.

First, it practically tells you how to get an idea for
it says that getting an idea is like creating a recipe for
a new dish. All you have to do is take some ingredients
you already know about and combine them in a new
way. It's as simple as that.

Not only is it simple, it doesn't take a genius to do

it. Nor does it take a rocket scientist or a Nobel Prize winner or a world-famous artist or a poet laureate or an advertising hotshot or a Pulitzer Prize winner or a first-class inventor.

"To my mind," wrote the scientist and philosopher Jacob Bronowski, "it is a mistake to think of creative activity as something unusual."

Ordinary people get good ideas everyday. Every day they create and invent and discover things. Every day they figure out different ways to repair cars and sinks and doors, to fix dinners, to increase sales, to save money, to teach their children, to reduce costs, to increase production, to write memos and proposals, to make things better or easier or cheaper—the list goes on and on.

Second, I like it because it zeros in on what I believe is the key to getting ideas, namely, combining things. Indeed, everything I've ever read about ideas talks about combining or linkage or juxtaposition or synthesis or association.

"It is obvious," wrote Jacques Hadamard, "that invention or discovery, be it in mathematics or anywhere else, takes place by combining ideas. . . . The Latin verb *cogito*, for 'to think,' etymologically means 'to shake together.' St. Augustine had already noticed that and had observed that *intelligo* means 'to select among.'"

"When a poet's mind is perfectly equipped for its work," wrote T. S. Eliot, "it is constantly amalgamating disparate experiences. The ordinary man's experience is chaotic, irregular, fragmentary. The latter falls in love or reads Spinoza, and these two experiences have nothing to do with each other, or with the noise of the typewriter or the smell of cooking; in the mind of the poet these experiences are always forming new wholes."

"A man becomes creative," wrote Bronowski, "whether he is an artist or a scientist, when he finds a new unity in the variety of nature. He does so by finding a likeness between things which were not thought alike before. . . . The creative mind is a mind that looks for unexpected likenesses."

Or listen to Robert Frost: "What is an idea? If you remember only one thing I've said, remember that an idea is a feat of association."

Or Francis H. Cartier: "There is only one way in which a person acquires a new idea: by the combination or association of two or more ideas he already has into a new juxtaposition in such a manner as to discover a relationship among them of which he was not previously aware."

Nicholas Negroponte agrees: "Where do good new ideas come from? That's simple—from differences. Creativity comes from unlikely juxtapositions."

And Arthur Koestler wrote an entire book, *The Act of Creation*, based on "the thesis that creative originality does not mean creating or originating a system of ideas out of nothing but rather out of a combination of well-established patterns of thought—by a process of cross-fertilization." Koestler calls this process "bisociation."

"The creative act," he explained, ". . . uncovers, selects, reshuffles, combines, synthesizes already existing facts, ideas, faculties, skills."

"Feats of association," "unexpected likenesses," "new wholes," "shake together" then "select among," "new (or unlikely) juxtapositions," "bisociations"—however they phrase it, they're all saying pretty much what James Webb Young said:

> An idea is nothing more nor less
> than a new combination of old elements.

Now that we know what ideas are, we must devise a method for getting them.

Happily enough, many such methods have already been devised. And—even more happily—these methods are quite similar.

In *A Technique for Producing Ideas*, James Webb Young describes a five-step method for producing ideas.

First, the mind must "gather its raw materials." In advertising, these materials include "specific knowledge about products and people [and] general knowledge about life and events."

Second, the mind goes through a "process of masticating those materials."

Third, "You drop the whole subject and put the problem out of your mind as completely as you can."

Fourth, "Out of nowhere the idea will appear."

Fifth, you "take your little newborn idea out into the world of reality" and see how it fares.

Hermann von Helmholtz, the German philosopher, said he used three steps to get new thoughts.

The first was "Preparation," the time during which he investigated the problem "in all directions" (Young's second step).

The second was "Incubation," when he didn't think consciously about the problem at all (Young's third step).

The third was "Illumination," when "happy ideas come unexpectedly without effort, like an inspiration" (Young's fourth step).

Moshe F. Rubinstein, a specialist in scientific problem solving at the University of California, says that there are four distinct stages to problem solving.

Stage one: Preparation. You go over the elements

of the problem and study their relationships (Young's first and second steps).

Stage two: Incubation. Unless you've been able to solve the problem quickly, you sleep on it. You may be frustrated at this stage because you haven't been able to find an answer and don't see how you're going to (Young's third step).

Stage three: Inspiration. You feel a spark of excitement as a solution, or a possible path to one, suddenly appears (Young's fourth step).

Stage four: Verification. You check the solution to see if it really works (Young's fifth step).

In *Predator of the Universe: The Human Mind*, Charles S. Wakefield says there "is a series of [five] mental stages that identify the creative act."

First, "is an awareness of the problem."

Second, "comes a defining of the problem."

Third, "comes a saturation in the problem and the factual data surrounding it" (Young's first and second steps).

Fourth, "comes the period of incubation and surface calm" (Young's third step).

Fifth, comes "the explosion—the mental insight, the sudden leap beyond logic, beyond the usual stepping-stones to normal solutions" (Young's fourth step).

Ah, but even though they all generally agree on the steps you must take to get an idea, none of them talks much about the condition you must be in to climb those steps. And if you're not in condition, it doesn't make any difference if you know the steps; you'll never get the ideas that you're capable of getting.

For telling most people how to get an idea is a little like telling a first grader to find x when $x + 9 = 2x + 4$, or like telling a person with weak legs how to high jump. Just as you must know algebra before you can solve an equation, and just as you must have strong legs before you can high jump, so you must condition your mind before you can get an idea.

The first ten chapters make up Part I of this book. They discuss *Ten Ways to Idea-Condition Your Mind*. You may read them in any order.

1. Have Fun
2. Be More Like a Child
3. Become Idea-Prone
4. Visualize Success
5. Rejoice in Failure
6. Get More Inputs
7. Screw Up Your Courage
8. Team Up with Energy
9. Rethink Your Thinking
10. Learn How to Combine

The last five chapters make up Part II of this book. They talk about *A Five-Step Method for Producing Ideas* that *should* be taken in sequence. Although I use different words, I too generally agree with Young. (Two exceptions: I add one step to his—the need to define the problem; and I combine his third and fourth steps because they seem one step to me, not two.)

To some, my (and Young's) last step may not seem part of the process of getting an idea, but it truly is, for an idea is not an idea until something happens with it.

11. Define the Problem
12. Gather the Information
13. Search for the Idea
14. Forget about It
15. Put the Idea into Action

Part I

Ten Ways to Idea-
Condition Your Mind

1.

Have Fun

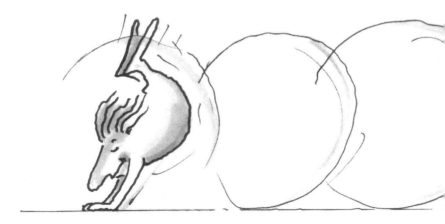

He who laughs, lasts.

Mary Pettibone Poole

Sometimes when reading Goethe I have the paralyzing suspicion that he is trying to be funny.

Guy Davenport

Seriousness is the only refuge of the shallow.

Oscar Wilde

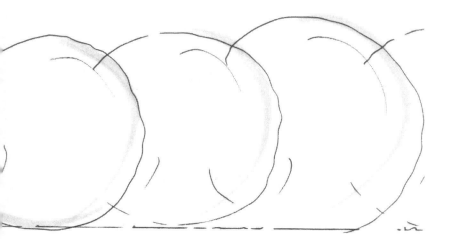

15

It's not by chance that I list having fun as my first suggestion on how to get your mind into idea-condition. Indeed, in my experience it might well be the most important one.

Here's why:

Usually in creative departments of advertising agencies a writer and an art director work together as a team on a project. In some departments and occasionally in the ones that I headed, three or four teams work on the same project.

When that happened in my departments, I always knew which team would come up with the best ideas, the best ads, the best television commercials, the best billboards.

It was the team that was having the most fun.

The ones with frowns and furrowed brows rarely got anything good.

The ones smiling and laughing almost always did.

Were they enjoying themselves because they were coming up with ideas? Or were they coming up with ideas because they were enjoying themselves?

The latter. No question about it.

After all, you know it's true with everything else—people who enjoy what they're doing, do it better. So why wouldn't it be true with people who have to come up with ideas?

"Make it fun to work at your agency," said David Ogilvy, the head of an advertising agency. "When people aren't having any fun they seldom produce good advertising."

Mr. Ogilvy did not have to limit his remarks to people in advertising agencies. The same could be said about anybody at any place who has to come up with an idea.

Oh, I know that creating advertising is a minor creative endeavor, and you might consider it folly to apply the lessons learned there to more weighty occupations. But people in other fields say the same thing about fun.

"Necessity may be the mother of invention," said Roger von Oech, "but play is certainly the father."

"Serious people have few ideas," said Paul Valéry. "People with ideas are never serious."

"The most exciting phrase to hear in science," said author and biochemist Isaac Asimov, "the one that heralds new discoveries, is not 'Eureka!' (I found it!), but 'That's funny . . .' "

Indeed, it should come as no surprise that humor and all kinds of creativity are bedfellows.

After all, as Arthur Koestler pointed out, the basis of humor is also the basis of creativity—the unexpected joining of dissimilar elements to form a new whole that actually makes sense; the sudden left turn when you were expecting the road to go straight; a "bisociation" (as Koestler puts it), two frames of reference slamming together.

Just listen to how it works in humor:

"How can I believe in God," asked Woody Allen, "when just last week I got my tongue caught in the roller of an electric typewriter?"

"The race may not be to the swift nor the victory to the strong," said Damon Runyon, "but that's the way to bet."

"Shut up, he explained," wrote Ring Lardner.

In every case your mind is going one way when suddenly you are forced to change directions and—wonder of wonders—this new, unanticipated direction is perfectly logical. Something new is created, something that after the fact often seems obvious.

Ah, but that's exactly what an idea is too. The unexpected joining of two "old elements" to create a new whole that makes sense, "two matrices of thought" (as Koestler puts it) meeting at the pass.

Johannes Gutenberg put a coin punch and a wine press together and got a printing press.

Salvador Dalí put dreams and art together and got surrealism.

Someone put fire and food together and got cooking.

Sir Isaac Newton put the tides and the fall of an apple together and got gravity.

Charles Darwin put human disasters and the proliferation of species together and got natural selection.

Levi Hutchins put an alarm and a clock together and got an alarm clock.

Hyman L. Lipman put a pencil and an eraser together and got a pencil with an eraser.

Someone put a rag and a stick together and got a mop.

I once went for a job interview to an advertising agency in Chicago. As soon as I walked in I knew it would be a good place to work, a place where ideas would be bouncing off the ceiling. As I got off the elevator, there on the wall was this big official-looking framed sign:

IN CASE OF EMERGENCY

1. Grab your coat
2. Get your hat
3. Leave your worries on the doorstep
4. Direct your feet to the sunny side of the street

There they were framed and hanging on the wall—"two matrices of thought" meeting at the pass, two frames of reference slamming together. Humor and creativity. It's difficult to have one without the other. The same is true for fun and ideas. And for enjoyment and performance.

Let me tell you a story:

When I started in advertising the writers and art directors dressed the way everybody in business dressed—the men wore suits and ties; the women, dresses or suits.

In the late sixties all that changed. People started dressing in sweaters and blue jeans and T-shirts and tennis shoes. I was running a creative department then and the *Los Angeles Times* asked me what I thought about people coming to work like that.

"I don't care if they come to work in their pajamas," I said, "as long as they get the work out."

Sure enough, the day after the article (with my quote) appeared, my entire department showed up in pajamas. It was great fun. The office rocked with laughter and joy.

More important, that day and the weeks that followed were some of the most productive times my department ever had. People were having fun, and the work got better.

Note again the cause and effect relationship: The fun came first; the better work, second. Having fun unleashes creativity. It is one of the seeds you plant to get ideas.

Realizing that, we started planting more of those seeds to make it fun to come to work. Perhaps a couple of them might work in your place, or will spark an idea for one that will work.

Meet in the Park. Our office was across the street from a park. Once a month or so we'd hold a department meeting there. (It's amazing how simply getting out of the office improved camaraderie and productivity.)

Family Day. Once a year, the kids came to see where mom and dad worked.

Darts. We put up a dart board in our conference room and played darts when we needed a break.

Who Is That? People brought in pictures of themselves when they were babies. We tacked all the pictures on a wall, numbered them, and everybody tried to guess who was who. The person who got the most right won a prize.

Cute/Homely Baby. Same as above, only we'd all vote on which baby was the cutest, which was the homeliest. Prizes, of course.

Arts and Crafts Fair. People sold (or just exhibited) things they or their families made at home.

Hallway Hockey. During lunch hour, we sometimes played hockey in the hallways with real hockey sticks, but with wads of paper for the puck.

Children's Art. Parents brought in their children's art work, labeled it, and hung it in the lobby.

Chili Off. The cooks in the department brought in pots of chili; we'd taste them and vote on a winner.

Dress-up Day. Every now and then we'd all come in dressed to the nines.

Potluck. Everybody brought in something, and we all sat down in the hallways and had lunch together.

"If it isn't fun, why do it?" says Jerry Greenfield of Ben & Jerry's Ice Cream.

Tom J. Peters agrees: "The number one premise in business is that it need not be boring or dull," he wrote. "It ought to be fun. If it's not fun, you're wasting your life."

Don't waste yours. Have some fun.

And not so incidentally, come up with some ideas.

2.

Be More Like a Child

A child is a curly, dimpled lunatic.

Ralph Waldo Emerson

There are more bores around than when I was a boy.

Fred Allen

Youth is such a wonderful thing. What a crime to waste it on children.

George Bernard Shaw

Insanity is hereditary—you get it from your children.

Sam Levenson

Charles Baudelaire described genius as childhood recovered at will.

He was saying that if you can revisit the wonder of childhood you can taste genius.

And he was right; it is the child in you who is creative, not the adult.

The adult in you wears a belt and suspenders and looks both ways before crossing the road.

The child in you goes barefoot and plays in the street.

The adult punches the ball to right.

The child swings for the fences.

The adult thinks too much and has too much scar tissue and is manacled by too much knowledge and by too many boundaries and rules and assumptions and preconceptions.

In short, the adult is a poop. A handcuffed poop.

The child is innocent and free and does not know what he cannot or should not do. He sees the world as it actually is, not the way we adults have been taught to believe that it is.

"In physics, as elsewhere," wrote Gary Zukav in *The Dancing Wu Li Masters*, "those who most have felt the exhilaration of the creative process are those who best have slipped the bonds of the known and venture far into the unexplored territory which lies beyond the barrier of the obvious. This type of person has two characteristics. The first is a childlike ability to see the world as it is, and not as it appears according to what we know about it."*

Mr. Zukav continued:

"The child in us is always naive, innocent in the simplistic sense. A Zen story tells of Nan-in, a Japanese master during the Meiji era who received a university professor. The professor came to inquire about Zen. Nan-in served tea. He poured his visitor's cup full, and then kept on pouring. The professor watched the overflow until he no longer could restrain himself.

" 'It is overfull. No more will go in!'

" 'Like this cup,' Nan-in said, 'you are full of your own opinions and speculations. How can I show you Zen unless you first empty your cup?' "

Mr. Zukav then said: "Our cup is usually filled to the brim with the 'obvious,' 'common sense,' and 'the self-evident.' "

"If you would be more creative," wrote the psychologist Jean Piaget, "stay in part a child, with the creativity and invention that characterizes children before they are deformed by adult society."

---

*"The second characteristic of true artists and true scientists," Mr. Zukav says later, "is the firm confidence which both of them have in themselves." (For a discussion of this characteristic, see chapter 3.)

J. Robert Oppenheimer agreed: "There are children playing in the streets who could solve some of my top problems in physics, because they have modes of sensory perception that I lost long ago."

Thomas Edison agreed too: "The greatest invention in the world is the mind of a child."

So did Will Durant: ". . . the child knows as much of cosmic truth as Einstein did in the ecstasy of his final formula."

Which is curiously close to what Albert Einstein himself said: "I sometimes ask myself how it came about that I was the one to develop the theory of relativity. The reason, I think, is that a normal adult never stops to think about problems of time and space. These are things that he has thought of as a child. But my intellectual development was retarded, as a result of which I began to wonder about space and time only when I had already grown up."

Perhaps Dylan Thomas put it best, though, when he wrote:

> The ball I threw while playing in the park
> Has not yet reached the ground.

Adults don't play in the park; children do.

Adults tend to do what they or other people did the last time.

To children there is no last time. Every time is the first time. And so when they go exploring for ideas they explore a land that is fresh and original, a land without rules, a land without borders or fences or walls or boundaries, a land infinite with promise and opportunity.

Remember the story in Robert Pirsig's *Zen and the Art of Motorcycle Maintenance* about the girl who couldn't think of anything to say when asked to write a 500-word essay about the United States? The teacher said to write about Bozeman, Montana, the town where the school was, instead of the entire United States. Nothing. Then he said to write about the main street in Bozeman. Still nothing.

Then he said, "Narrow it down to the front of one building on the main street of Bozeman. The Opera House. Start with the upper left-hand brick."

The next class the girl turned in a 5,000-word essay on the front of the Opera House on the main street of Bozeman.

" 'I sat in the hamburger stand across the street,' she said, 'and started writing about the first brick, and the second brick, and then by the third brick it all started to come and I couldn't stop.' "

She was initially blocked, wrote Pirsig, "because she was trying to repeat, in writing, things she had already heard. . . . She couldn't think of anything to write about Bozeman because she couldn't recall anything she had heard worth repeating. She was strangely unaware that she could look and see freshly for herself, as she wrote, without primary regard for what had been said before."

Children don't have such blockages because children don't know about before. They only know about now. And so when searching for a solution to a problem they look and see freshly for themselves. Every time.

They break rules because they do not know the rules exist. They do odd things that make their adult parents uneasy. They stand up in the boat and rock it. They shout in church, play with matches, and pound the piano with their fists.

They constantly see the new relationships among seemingly unrelated things. They paint trees orange and grass purple, and they hang fire trucks from clouds.

They study ordinary things intently—a blade of grass, a spoon, a face—and have a sense of wonder about the things that most of us take for granted.

They ask and ask and ask.

"Kids are natural-born scientists," said Carl Sagan. "First of all, they ask the deep scientific questions: Why is the moon round? Why is the sky blue? What's a dream? Why do we have toes? What's the birthday of the world? By the time they get into high school, they hardly ever ask questions like that."

"Children enter school as question marks and leave as periods," agreed Neil Postman.

Become a question mark again.

Whatever you see, ask yourself why it is the way it is. If you don't get an answer that makes sense, perhaps there's room for improvement.

Why is your production line set up the way it is?

Why does your receptionist sit behind a desk? Why do you?

Why do you come to work and leave when you do? Why does your office or plant open and close when it does?

Why do your business cards, your stationery, your presentation books all look the way they do?

Why does your product look the way it does?

Why is your product packaged the way it is?

Why do your bills and invoices look the way they do?

Why are kitchen counters and bathroom sinks the height that they are?

Why don't kitchen faucets have foot pedals?

Why don't refrigerators have pull-out drawers?

Lots of banks have their customers form one long line so that no customer ever gets stuck in a slow line. Why don't supermarkets and other stores do that?

Why is the word "Milk" so often the biggest or the second biggest word on milk cartons? Everybody knows it's milk. Why isn't that space put to better use?

Why don't they put gasoline caps on both sides of your car, so that no matter which side of the gasoline pump you park on, you'll never have to pull the hose around to the other side?

All of us have mental pictures of ourselves. How old is the person you see in your mental image? When I put that question to one of the most creative people I know (the illustrator of this book), he said, "Six."

Imagine. When he thinks of himself, he thinks of a six-year-old.

No wonder he continually comes up with fresh solutions and ideas. He unconsciously thinks like a six-year-old much of the time, seeing things through a six-year-old's eyes.

Once when we were working on a cat food commercial he wondered what the world looked like from a cat's point of view—when the cat was running, what did the walls and stairs and furniture look like to it? What did it dream about? What did its food look like? Did its canned "Salmon Dinner" look to it like a salmon dinner looks to us? The questions went on and on.

Another almost-as-young friend of mine was working on a Smokey Bear commercial and wondered what it would be like if the animals in the forest came into our backyards every summer and left their still-smoldering campfires burning when they left, in the same way that we leave our fires still smoldering in their backyards.

Still another friend wondered what it was like after hours in the produce department of a grocery store. Did the Sunkist lemons tell the broccoli they'd make a beautiful twosome?

Let the child in you come out. Don't be afraid.

Most businesses reward people who come up with new ideas. And one of the ways to come up with new ideas is to be more like a child.

So next time you have a problem to solve or an idea to come up with, ask yourself: "How would I solve this if I were six years old?" "How would I look at this if I were four?"

Loosen up. Run down the hall someday at work. Eat an ice cream cone at your desk. Take everything out of your desk drawers and put it on the floor for a couple of days. Rearrange your office furniture. Take a nap after lunch. Draw pictures on your window with a felt-tip pen. Write notes with crayons. Sing out loud in the elevator. Pound the piano. Stand up in the boat. And then rock it.

Have fun. (See chapter 1.)

Forget what was done before. Break the rules. Be illogical. Be silly. Be free.

Be a child.

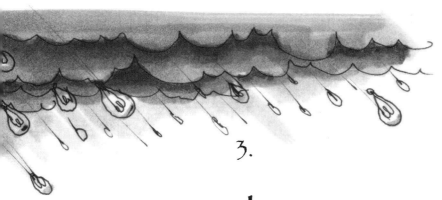

3.

Become Idea-Prone

That fellow seems to me to possess but one idea, and that a wrong one.

Samuel Johnson

Everyone is a genius at least once a year; a real genius has his original ideas closer together.

G. C. Lichtenberg

Man can live without air for a few minutes, without water for about two weeks, without food for about two months—and without a new thought for years on end.

Kent Ruth

Nobody understands (yet) how your brain—a physical thing—can produce an idea—something that is not physical.

All we know is that it happens. Perhaps it happens to you less often than it does to other people. But since it has happened to you a few times we know that there's no physical deficiency—no genetic mutation in your brain, for example—that's preventing you from getting ideas. You can get them. That's proven.

So the only thing we have to figure out is why you're getting too few of them and then work on getting more of them.

When I was a kid I hung around with a guy named Johnny-Boy Boyd. JB was a klutz. Accidents just seemed to happen to him; if one didn't run into him, he ran into it.

Nowadays psychologists would say that subconsciously JB made accidents happen, that it was his way of getting attention.

Back then we just called him "accident-prone" and let it go at that.

As an adult I hung around with people who were "idea-prone." Ideas just seemed to happen to them the

way accidents happened to JB. And the psychologists
would probably say the same thing about them that
they said about JB—that subconsciously they made
these things happen, that it was their way of getting
attention.
Perhaps. But I think there's more to it than that.

James Webb Young quotes Vilfredo Pareto, who
believed that there were two main types of people,
types he called the *Speculator* and the *Rentier*.
The *Speculator,* said Pareto, is constantly
preoccupied with the possibilities of new
combinations. This is the type that includes, as Young
put it, "all those persons in any field who . . . cannot
let well enough alone and who speculate on how to
change it."
The *Rentier,* on the other hand, includes "routine,
steady-going, unimaginative, conserving people whom
the *Speculator* manipulates."
Young agreed with Pareto that those two types
exist and—that being the case—concluded "that there
are large numbers of people whom no technique for
producing ideas will ever help."
I disagree with that conclusion.
I do not think that the idea-prone people I hung
around with were born with some special kind of
ideation talent, or some unique way of thinking that
led them down untrodden paths, or some laserlike
insight that let them see order and new relationships
where others saw only chaos.

The thing that sets them apart is this:

The ones who come up with ideas know that ideas exist and know that they will find those ideas; the ones who don't come up with ideas don't know that ideas exist and don't know that they will find ideas.

Let me say it again:

The ones who come up with ideas know that ideas exist and know that they will find those ideas; the ones who don't come up with ideas don't know that ideas exist and don't know that they will find ideas.

Know That Ideas Exist

When I first started teaching I told my students that for every problem there was a solution, an answer, an idea.

I was wrong.

I now know that there are hundreds of solutions, hundreds of answers, hundreds of ideas.

Maybe even thousands. Indeed, perhaps it's infinite.

Just consider:

As of 1940 (the last time they checked, I guess) a total of 94 patents had been taken out on shaving mugs. Shaving mugs, for heaven's sake!

There are more than 1,200 different kinds of barbed wire.

Enough cookbooks have been published in the United States to fill a small library.

Or just listen to Lincoln Steffens, writing in 1931:

> Nothing is done. Everything in the world remains to be done or done over. The greatest picture is not yet painted, the greatest play isn't written, the greatest poem is unsung. There isn't in all the world a perfect railroad, nor a good government, nor a sound law. Physics, mathematics, and especially the most advanced and exact of the sciences, are being fundamentally revised. Chemistry is just becoming a science; psychology, economics, and sociology are awaiting a Darwin, whose work in turn is awaiting an Einstein.
>
> If the rah-rah boys in our colleges could be told this, they might not all be specialists in football, parties, and unearned degrees. They are not told it, however; they are told to learn what is known. This is nothing.

Every word he wrote is as true today as it was in 1931. Nothing is done. Everything waits for you to do it.

Let me tell you a story:

For more than 20 years I worked for the advertising agency that did the advertising for Smokey Bear. The first thing the writers and art directors had to do every year was to come up with a basic poster.

The rules for the poster never varied: It had to be a certain shape and size; it had to feature Smokey; it had to be simple enough to grasp at a glance, clear enough for even a dunce to understand, and (if it had words) brief enough to read in three or four seconds.

The mission of the poster never varied either: It had to convince people to be careful with fire.

In other words, every year we had to come up with the same thing only different.

And we did. Indeed, every year we came up with 15 or 20 different ideas for posters. Every year. For more than 20 years. Over 300 posters all featuring Smokey and all trying to do the same thing and not two of them the same.

As far as I know, the writers and art directors at that agency still have the same rules and mission for the Smokey poster and are still coming up with ideas.

So don't tell me there's only one or two ways to solve a problem. I know differently.

Or listen to a story a friend of mine told me:

"I used to teach a three-day seminar on advertising in Chicago. One of the assignments I gave each student was to create, overnight, an outdoor board for a Swiss Army knife. Most of the students would come in the next morning with the required billboard, but several of them would say that they worked for hours and couldn't come up with anything. This happened three years in a row.

"The fourth year I tried something different. Instead of asking for just one billboard, I asked each student to create at least ten billboards for a Swiss Army knife. And instead of giving them all night, I told them they had to do it during their lunch hour.

"After lunch everybody had at least ten ideas. Many had more. One student had 25.

"I came to realize that when faced with a problem most people look for the one right solution because that's the way they were brought up. All through school they had to answer multiple-choice and true-or-false questions, questions that had only one right answer. And so they assume that all questions and problems are like that. And when they can't find a solution that looks perfect they give up.

"But most problems aren't like exam questions in school. Most problems have many solutions. And as soon as I forced my students to realize that, they found those solutions."

Did you hear that? As soon as his students realized there were many solutions they found those solutions.

"Always think of what you have to do as easy and it will become so," said Émile Coué.

When you're not sure an answer exists, finding it can be hard. When you know there are many answers, finding one or two is easy.

Dr. Norbert Wiener noticed the same thing: "Once a scientist attacks a problem which he knows to have a solution, his entire attitude is changed. He is already some 50% of his way toward that answer."

Arthur Koestler agreed: "The mere knowledge that a problem is soluble means that half the game is already won."

That's one of the reasons some people always seem to get ideas—they know they're around.

I was working one day in my office with Larry Corby, the illustrator of this book, trying to come up with TV commercials for a kid's toy.

"Shut the door," he said.

"Why?"

"There are a couple of ideas in here and I don't want them to get out."

He wasn't joking either. He truly believed that the ideas we were looking for were physically present in that room. And since he knew they were there, five minutes later he found a couple of them.

Joseph Heller believed the same thing. "I feel that these ideas are floating around in the air and they pick me to settle upon," he said.

And so of course did Thomas Edison. He believed—no, he knew—that ideas "are in the air." If he hadn't found them someone else would have. Is it any wonder he was the one who found so many?

There's always another idea, always another solution. Accept it.

Know That You Will Find Those Ideas

You now know (I hope) that hundreds of solutions exist for your problem, that ideas abound. OK then, why aren't you finding them?

Consider these three events:

1. You've seen this one happen all the time: Some golfer you've never heard of is leading a major tournament after the first day. The newspapers are full of stories about him. Everybody's talking about him. He's the new Arnold Palmer, the new Jack Nicklaus, the new Tiger Woods.

The next day the poor guy shoots eight-over, misses the cut, and disappears.

What happened?

2. I was filming a commercial at the Los Angeles Forum once and Wilt Chamberlain was at the other end of the court practicing free throws. He had a couple of kids there retrieving the balls for him. He must have shot over 100 free throws while I was there and I saw him miss only three. Swish, swish, swish. It was remarkable. That night in a game he missed eight out of twelve from the line.

What happened?

3. You have a speech to deliver out of town and you've got it down cold. You know your subject, you know what you want to say, you know how you want to say it. A piece of cake. You rehearse it in front of a mirror. A perfect 10.

But when you get up to deliver it your mind goes blank and the speech is a disaster.

What happened?

You know what happened.

There are a lot of different ways to say it, but basically you and Wilt and that forgotten golfer—consciously or unconsciously—all started doubting yourselves. And the rest is history.

The golfer on that first day and Wilt on the practice court and you in your hotel room all felt comfortable at the levels you were performing at.

But later you somehow got to wondering if you were as good as you thought you were. Your performances on the golf course and on the practice court and in the hotel room were better than the mental images you had of yourselves.

So your bodies and minds automatically lowered your performances to the levels where you felt comfortable again.

And no amount of will power, no amount of effort or practice or determination could bring your performances back to where they were.

That's because your self-image determines what you are and how you perform. Not effort or will. Self-image.

And the only way to significantly improve your performance is to improve your self-image.

So if you want to become idea-prone you must accept two things:

First, you must accept that what you think about yourself is the single most important factor in your success.

Your personality, your actions, how you get along with others, how you perform at work, your feelings, your beliefs, your dedication, your aspirations, even your talents and abilities are affected—no, controlled— by your self-image.

You act like the kind of person you imagine yourself to be. It's as simple as that.

And it's no longer open to question.

If you think of yourself as a failure you will probably become a failure. If you think of yourself as successful you will probably become successful.

How else can you explain why seemingly gifted people fail while seemingly deprived people succeed?

"They can do it all because they think they can," said Virgil, and this fundamental fact about the triumph of self-image is as true today as it was 2,000 years ago.

Henry Ford agreed: "Whether you think you can or can't, you're right."

In short: Attitude is more important than facts.

Specifically this means that for the most part the difference between people who crackle with ideas and those who don't has little to do with some innate ability to come up with ideas. It has to do with the belief that they can come up with ideas.

Those who believe they can, can; those who believe they can't, can't. It's as simple as that.

Second, you must accept that what William James called "the greatest discovery of my generation" is also a fact. The discovery?

Human beings can alter their lives by altering their attitudes.

Jean-Paul Sartre put it like this: "Man is what he conceives himself to be."

And Anton Chekhov put it like this: "Man is what he believes."

This too is no longer open to question.

And yet this is what many people, perhaps yourself included, refuse to accept.

You accept that your self-image drives your life, but despite all the evidence cited by sages and parents and clergymen and doctors and poets and researchers and philosophers and psychologists and teachers and therapists and coaches, and despite the thousands of real-life examples in the hundreds of self-improvement books, you reject the notion that you can change your own self-image.

You are wrong. You can change it.

You accept "As a man thinketh in his heart so is he." But you seem to believe that if you thinketh differently in your heart you will remain the same you.

You won't. You will be a different you.

Or you seem to think that you can't think differently in your heart, that the way you think today is locked in stone forever.

You are wrong. You can think differently.

Everybody accepts now that the mind alters how the body works. The evidence that it can and does is simply overwhelming.

Drug addicts take placebos and have no withdrawal symptoms, allergy sufferers sneeze at plastic flowers, unloved children physically stop growing, hypnotized patients undergo surgery without anesthesia, people lower their blood pressures and pulse rates by willing it, cancer victims experience spontaneous remissions, hopeless cripples walk away cured from Lourdes—the examples are legion.

But when you think about it, accepting the concept that one thing (the mind) can alter another thing (the body) is a huge leap, a major leap, perhaps even a quantum leap.

All I'm asking you to accept is a minor leap—that the mind can alter the mind.

Accept it. It's a fact.

And then start altering your self-image.

I do not propose to tell you in this book how to do it, except to say this: If you tell yourself that you "never get ideas," you never will.

Instead, tell yourself every day that you are a font of ideas, that ideas bubble forth from you like water from a spring. Every day. No, many times every day. Eventually you will begin living up to this new mental image you have created of yourself.

Of course the libraries and bookstores are loaded with hundreds of books and tapes and videos that can tell you much better than I can how to change your self-image—*The Magic of Believing; Change Your Life Now; Psycho-Cybernetics; Think and Grow Rich; The Power of Positive Thinking; Life's Too Short; Unlimited Power*—the list goes on and on.

Get one of them and read it.

Every one says basically the same thing—that you can change your life by changing the way you think about yourself.

And every one of them is right.

Accept it.

Once you know that ideas exist and that you will find them, a great calm envelops you. It is a calm you need today more than ever.

The reason?

Today wasn't supposed to break this way.

Computers and faxes and modems and e-mail and the Internet and voice mail and cell phones and networking were all supposed to make our lives simpler and easier. We were supposed to have more time than ever to come up with ideas.

But for many—perhaps for you—the reverse happened. Downsizing stole the time that electronics created. And now it seems you have less time to do twice as much. And that squeeze is starting to panic you.

Well, relax. You know the idea is out there. And you know that you're going to find it.

So don't worry about time. Although some ideas take longer to get than others, getting an idea does not depend, strangely enough, upon time. Nor upon workplaces or schedules or even workloads.

You can search for an idea while you're eating lunch or taking a shower or walking your dog. And you can find it in the instant you start your car or snap on a light.

Getting an idea depends upon your belief in its existence. And upon your belief in yourself.

Believe.

4.

Visualize Success

The brain is a wonderful organ. It starts working the moment you get up in the morning and does not stop until you get to the office.

Robert Frost

The trouble with the rat race is that even if you win, you're still a rat.

Lily Tomlin

When I go to the beauty parlor, I always use the emergency entrance. Sometimes I just go for an estimate.

Phyllis Diller

I want you to imagine a steel beam about one foot wide and one hundred feet long.

Let's say I take that steel beam up to the top of a forty-story office building and lay it across to the top of a forty-story office building on the other side of the street.

Now here's the deal: If you walk across that beam from one building to the next I'll give you $100.

If you're like most people you'll say forget it. "Walk across that narrow piece of steel forty floors up? No way. I could lose my balance and fall." And you probably would have too.

Now I go across the street to the other building and hold your twelve-week-old baby girl over the side and tell you that unless you walk across that beam right now I'll drop her.

If you're like most people you'll walk across the beam. Not only that, you'll probably make it easily, walking across it as effortlessly as you walk across a bridge.

Why did you react so differently? The task— walking across the beam—didn't change.

You reacted differently because the goals you were visualizing changed.

The first time your goal was not to fall.

The second time your goal was to save your baby.

The first time you were concerned about the getting there—how you should place your feet, how you should hold your arms for balance, how fast you should go, how long your stride should be, how you should keep from falling.

The second time you didn't think about any of those things. All you thought about, all you visualized, was saving your baby. And your mind automatically figured out how your body should move in order to get there.

In the same way, if you set your mind on goals— on getting ideas, for example—your mind will figure out a way to get them.

Or ponder the case of the guy who was trying to develop a computer program that would determine where and when and how fast a center fielder should run when a baseball was hit in order to catch it like Willie Mays.

He had to consider the wind and the humidity at the ball park, the sound of the bat hitting the ball, the kind of pitch the pitcher threw, what that particular batter had done in previous situations against that particular pitcher and that particular pitch in that particular ballpark, and how that particular batter had been hitting the ball lately.

He had to consider the speed of the ball as it left the bat and how that speed would decrease the farther it went.

He had to consider the direction and rotation of the ball and the angle of its rise and descent.

Then he had to consider how fast the fielder should run, and in what direction and at what angle, in order to catch the ball before it hit the ground or the wall.

I don't know if he succeeded in developing such a program.

But I do know that Willie Mays did all that without consciously thinking about any of it.

He just saw the ball being hit and ran to the precise spot on the field where the ball was going. All he visualized was the goal—catching the ball. His brain took all the information that his eyes and ears and memory were furnishing and did all the computing for him: It told his body where to go, his legs how fast to run, his arm how high to reach, his hand which angle to turn.

Let me give you another example:

Research Quarterly reported on a study that showed how practicing basketball free throws in your mind affects your performance.

One group of students actually practiced shooting free throws every day for twenty days, and each student was scored on the first and last day.

Each student in the second group was also scored on the first and last day, but they did no practicing in between.

Students in the third group imagined shooting free throws every day for twenty days, mentally correcting their errors when the ball didn't go in; they were also scored on the first and last day.

Students in the first group—those who actually practiced—improved their shooting by 24 percent.

Students in the second group—those who did nothing—showed no improvement.

And students in the third group—those who practiced in their imaginations by visualizing— improved their shooting by 23 percent.

Experiments with dart throwers showed the same thing—that mentally throwing darts at a target improves aim as much as physically throwing darts at it.

Case closed.

Because don't you see? Once again it's a quantum leap versus a minor leap situation.

If your mind can control the way your body behaves and acts on twelve inches of steel forty floors up, or on a baseball field when a ball is hit, or in front of a dart board, or on a basketball court—if your mind can control the way your body works to that extent, just think of the way your mind can control the way your mind works.

So if you want to get ideas, imagine having gotten them.

Visualize the scene the way the students visualized the ball going in the hoop, the darts into the target.

Visualize it the way divers visualize the dive, pool players the shot, tennis players the slam, golfers the putt.

Do not imagine that you will get the idea. Imagine that you already have it. Imagine being praised and thanked and rewarded.

There's a good chance you will be.

5.

Rejoice in Failure

It is not enough to succeed. Others must fail.

Gore Vidal

The theory seems to be that so long as a man is a failure he is one of God's chillun, but that as soon as he has any luck he owes it to the Devil.

H. L. Mencken

I picture my epitaph: "Here lies Paul Newman, who died a failure because his eyes turned brown."

Paul Newman

To swear off making mistakes is easy. All you have to do is swear off having ideas.

Leo Burnett

Here are five reasons why you should make failure a friend:

1. The only way to know that you've gone far enough is to go too far. And going too far is called failing.

But if you don't go far enough in searching for an idea—if you don't, in other words, fail—you can't be sure you've got the best idea.

So never fear failure or try to avoid it. Embrace it. Rejoice in it. It's a sign that you've gone far enough.

Race car drivers know this in their bones. They even have a saying about it:

"The one sure way to find out if you're going fast enough is to crash."

Cooks know it too.

Nearly everything they make has an "Oops, we've gone too far" point to it.

And the only way they can learn where that point is, is to go past it.

So they learn not to burn rye toast or Porterhouse steaks, not to oversauté chicken breasts or garlic or scallops, not to overmix whipping cream, or oversteam broccoli, or overwhip egg whites, not to overbake pork roasts or cakes or soufflés, not to . . .

The lessons are endless.

And the only way they can learn them irrevocably is by failing.

Emulate race car drivers and cooks.

Go too fast. Go too far. Let your mind wander into dangerous ground, into silliness, absurdity, stupidity, impossibility. Surprise your teachers. Astound your friends. Embarrass your parents. Thumb your nose at the laws of nature and science and common sense.

Crash.

Burn.

2. Ralph Price, an advertising agency art director I used to work with, says the same thing about failure. "You don't know if you've succeeded until you fail," he used to say. But he meant something slightly different.

He meant that many times you don't know if an idea is any good until you have other ideas to compare it to.

That's why writers and art directors in advertising agencies come up with many ideas on every project they're working on.

I suggest you do the same thing on the problem you're working on. Once you come up with an idea that seems to work, put it aside. It won't go anywhere.

Then come up with another idea. And then another one. And then . . .

For as I explain in chapter 13, there's always another one. Always.

3. "I have not failed," Thomas Edison said. "I've just found 10,000 ways that won't work."

Emulate Edison. Put a positive spin on things when they don't work out. Believe that every failure brings you one step closer to success. It will prevent you, as it did Edison, from becoming discouraged. More, it will urge you on.

4. When you fail, it changes your mind set, the way you look at life.

Failing makes you fearless. Failing sets you free.

Perhaps Jerry Della Femina, the famous advertising man, said it best:

> Failure is the mother of all creativity. My advice to anybody who wants to be creative is to get into something that will fail. I've failed at a lot of things in my life and I hope to fail at a lot more. Most people are afraid to fail, but once you've done it you find out it's not that terrible. There's a sense of freedom that you get from taking chances.

A friend of mine who was opening an office in Los Angeles for a major national advertising agency also knew this about failure. He was deluged with hundreds of applications.

"What kind of people are you looking to hire?" I asked him.

"The usual. Writers, art directors, account people, media, research—you know."

"But how do you choose, for example, one good media person from another?"

"I have to like them. If I don't like them, I won't hire them no matter how well qualified they are."

"What else?"

"I must confess I'm partial to people who have failed."

"What?"

"People who have failed. They know that failure's never permanent. Too often, people who haven't failed at anything think that failure's a disaster, and so they're afraid to go to the edge of what's possible. They're afraid to take chances. And because they've never failed, they think they know it all. I hate know-it-alls. Besides, you're always getting rejected in this business. That's just the way it is. I want people who I know will spring back."

The space program, so the story goes, felt this way too.

In selecting astronauts, legend has it that NASA looked for some evidence of failure in the resumes of the recruits. They knew that at some point in a space journey, the unexpected might happen—things might break down, fail, not go as planned.

They wanted people who would not be unnerved by such a situation; people who had experienced failure before and had learned from it, become wiser and stronger from it; people who knew that failure is nothing but a temporary setback, a prelude to success, a door being opened, as well as a door being closed.

So don't hide your failures or be ashamed of them.

Wear them with pride. Revel in them.

5. Of course what I've been talking about are failures where you know things aren't working—crash-and-burn failures, failures that point you in another direction, failures that teach you something.

Every now and then, however, you know in your heart that your idea is a good one, that your solution will work if given the chance, that what you've done is right.

When that happens, use the failure as a motivation to keep trying.

The "I'll show them!" drive is a powerful vehicle. Ride it.

There are hundreds of stories about how this kind of stubborn refusal to accept failure eventually led to success.

Chester F. Carlson, inventor of the Xerox machine, spent seventeen years trying to get companies interested in his photocopying device.

Bette Nesmith Graham made her Liquid Paper (then called "Mistake Out") in her kitchen for a decade before it started to sell big.

Alfred Mosher Butts aggressively marketed his Scrabble game for four years before it caught on.

It took James Russell twenty years to convince the music industry to adopt his "digital music" invention.

Catch 22 was turned down by 23 publishers; Dr. Seuss by 24; *Sister Carrie* by 28; *Chicken Soup for the Soul* by 33; *Zen and the Art of Motorcycle Maintenance* by 121.

Finally, if you'll allow me—a personal story of stubbornness:

Over a period of two years, I sent the original manuscript for this book to 74 publishers. Every time I got a rejection, I sent it out to a couple more publishers. The 44th one, Steven Piersanti of Berrett-Koehler, decided to publish it four months after I sent it to him. (While he was deciding, I sent it to 30 other publishers.)

Had I given up after getting 43 rejections, there would not have been a first edition of *How to Get Ideas*, a book that has sold nearly 100,000 copies and been translated into fifteen languages.

Nor would you be reading this second edition now.

6.

Get More Inputs

It is now proved beyond doubt that smoking is one of the leading causes of statistics.

Fletcher Knebel

Knowledge is power, if you know it about the right person.

Ethel Watts Mumford

We are here and it is now. Further than that all human knowledge is moonshine.

H. L. Mencken

Over the years I worked with hundreds of creative people in advertising agencies. These were people who got ideas for a living. On demand. Every day.

They came in all shapes and sizes and colors and signs and personalities. One had a doctorate in anthropology, one never got past third grade. They came from close families and from broken homes, from penthouses and from ghettos. I worked with gays and straights, with extroverts and introverts, with flashers, drunkards, suicidals, ex-priests, ex-touts—the list goes on and on.

But they all had two characteristics in common.

First, they were courageous, a subject I will deal with in the next chapter.

Second, they were all extremely curious. They had an almost insatiable curiosity about how things work and where things come from and what makes people tick.

They were curious about pie-making machines and flower drying, about Aztec wedding ceremonies and motorcycle design, about phobias and limes.

They knew things like the name of the horse Napoleon rode at Waterloo (Marengo), how many times egg whites increase in size when whipped (seven), how much a ten-gallon hat holds (three-quarters of a gallon), and the average number of times African elephants defecate every day (sixteen).

Most came by this curiosity naturally. All their lives they had, as one person put it to me, "the need to know." In some, this need was so overpowering that they came to think of it as a curse instead of a blessing. They were wrong.

For their curiosity was one of the reasons they were able to come up with ideas in the first place. Their curiosity was forcing them to continually accumulate bits of knowledge—"general knowledge about life and events"—the "old elements" that James Webb Young talked about.

And someday they'll combine those elements with other elements to create an idea. And the more elements they have to combine, the more ideas they can create.

After all, if "a new idea is nothing more nor less than a new combination of old elements," it stands to reason that the person who knows more old elements is more likely to come up with a new idea than a person who knows fewer old elements.

A copywriter named Jeff Weakley knew the importance of being curious. He sent me the best resume I ever got. It was in the form of a magazine ad.

The illustration showed a man's head piled high with junk—broken pencils, old tires, bottles, and so on.

The headline said: "Invest in a junk pile."

The copy said:

I read the labels on Vienna Sausage cans. I have worked as a porn projectionist, a disc jockey, and a door-to-door pot-holder salesman. I once read Freud and watched Laverne and Shirley at the same time. The leader of the King family was named Elmer. I have studied and worked with film, television, radio, and photography. I play sports, speak Spanish, and know the names of all the teeth in my mouth. I have read a lot of books and seen a lot of movies. I know of a disease that causes men to eat dirt, ice, and laundry starch. Most importantly, I know something about Egyptian burial customs and dual-in-line-package 14-pin quad two-input NANS gates.

I've always been like that. I've created in my mind this giant junk pile of seemingly worthless information.

And then I discovered advertising. The junk pile became a treasure overnight.

As I am at the earliest stages of my advertising career, I am offering you the opportunity to invest in this junk pile. For a very good price.

If you're interested in seeing the rest of my collection, please call.

Thank you.

If you do not have a natural curiosity like Jeff's that forces you to accumulate bits of knowledge, you must force yourself.

Every day. Deliberately.

Every day since he was twelve, Ray Bradbury once told me, he read at least one short story, one essay, and one poem. Every day. He said he never knows when something he read twenty years ago will "collide" (his word) with something he read yesterday to produce an idea for a story.

When was the last time you read a short story or an essay or a poem? Is it any wonder Ray Bradbury comes up with more story ideas than you do?

Here are two ways to force yourself to get more old elements:

1. Get Out of Your Rut

Of course you're in a rut. Admit it.

Why do you think you do the same things the same way in the same order every morning when you wake up? Or have the same thing for breakfast every day? Or go to work the same way every day? Or always read the same parts of the newspaper? Or always buy the same things at the supermarket? Or always watch the same TV programs? Or eat the way you eat, or dress the way you dress, or think the way you think, or, or, or?

It's because you're in a rut.

And because you're in a rut, every day your five senses are recording the same things they recorded yesterday—the same sights, the same feelings, the same smells, the same sounds, the same tastes.

Oh, sure, different things creep in every now and then. You can't help it. Not even a deaf, blind hermit can keep out new sensations.

But they creep in, in spite of what you're doing not because of what you're doing.

And if you just stay in your rut and let things creep in naturally, you'll never pile up the kind of varied and extensive database you need to form new ideas.

There's a huge, fascinating, exploding world of information out there—in any direction you care to look.

But you must look. And the sooner you do the sooner you'll become aware of "old elements" you didn't even know existed.

As Jerry Della Femina says: "Creativity is about making a lot of quick connections—about the things you know, the things you've seen. The more you've done, the easier it is to make that jump."

Perhaps that's why André Gide was so creative. Legend has it that he tried to read at least one book every month about a subject in which he had no interest. Have you ever done that? Do it. At least once.

Also:

Listen to a radio station you've never listened to before.

Study Latin.

Order something at a restaurant without knowing what it is exactly.

Read the want ads. Read Marianne Moore and Allen Ginsberg and Ted Kooser. Read a children's book. Reread *Death of a Salesman*. Read a magazine you've never heard of.

Check out something on the Internet you think you'll dislike. See a play or a movie you think you'll dislike. Rent a video you've never heard of.

Touch the bark of three different trees in your neighborhood. Learn to tell which is which simply by how it feels. Learn to tell which is which simply by how it smells.

Go to lunch with someone different.

Listen intently to music you don't like.

Ride the bus for a week.

Learn to read music. Learn sign language. Learn to make quenelles. Learn to tie knots.

Take up watercolor painting.

Study Greek. Or Chinese. Or English.

Visit a store, a gallery, a museum, a restaurant, a market, a mall, a building, a place you've never visited before.

Of course I'm not saying to do all of these things.

But please, today, do something. Something different, something that will get you off dead center, something that will start you in a different direction, something that will get you out of your rut.

"If you want to be creative," Louis L'Amour said, "go where your questions lead you. Do things. Have a wide variety of experiences."

A writer friend of mine in Los Angeles lived about ten miles from his office. It was a straight shot, right down Wilshire Boulevard from Westwood to downtown. But he never once took Wilshire. Indeed, every weekday morning for nine years he drove to work a different way. Never, he claims, did he go the same way twice. "Admittedly, there were times when I had to do some pretty crazy things to keep from taking the same route," he remembers. "I had to drive down alleys, and wander through residential areas, and get on freeways going away from where I wanted to go. But I never repeated myself. And I bet I saw more of Los Angeles in those nine years than most people will see in their lifetimes."

In what way are you richer than my friend because you go to work the same way every day?

Every day he saw something he never saw before. Every day you see the same things you saw yesterday. He was constantly seeing new things. You are constantly seeing the same old things.

Tomorrow go to work a different way. And a different way the next day. And the next. Forever.

2. Learn How to See

Back before the Second World War my dad and mom
and I used to drive from Evanston, Illinois, to my
mom's parents' house in Danville, Illinois. We went
every month or so. It was about a two- or three-hour
trip in those days. Sometimes we played a game
called "White Horse" on the way. It was a simple game
really—the first person to spot a white horse by the
side of the road or way back off in some pasture said,
"White Horse," and at the end of the trip the one who
spotted the most white horses first won the game.

Now the interesting thing I remember about that
game is that when we played it we saw all sorts of
white horses. But when we didn't play the game we
hardly saw any at all.

Why?

It wasn't because there were all sorts of white
horses around when we were playing and only a
couple around when we weren't.

It was because when we were looking for those
white horses we saw them; when we weren't, we didn't.

The same thing happens when you just buy a car,
or even if you're just thinking about buying a car. All
of a sudden you start seeing cars like that all over the
place.

They were always there before. You just didn't
see them because you weren't looking for them. But
as soon as you got interested in that particular car you
started—consciously or not—looking for them. And
voilà, there they were.

And what's true about white horses and cars is true about everything.

For you see everything that comes in contact with your eyes.

You see every car that you passed on the way to work this morning. And every car that passed you. And every driver in every car that you passed or that passed you.

You see every tree and bush and patch of grass you passed too. And every telephone pole, every gas station, every building, every traffic light, every person, every street lamp, every mailbox, every everything.

Then how come you can recall only a fraction of what you see?

It's because you weren't really seeing. You were simply looking. Not looking for, just looking. Looking requires no effort at all. It's as easy as breathing. Seeing is different; it requires effort. And commitment.

But hear this: Once you get the hang of it, seeing becomes almost as natural as looking.

Let me tell you a couple of stories:

Evanston, where I grew up, was dry. If you wanted to get a drink you had to go to Skokie or down to Howard Street, the street that divided Evanston from Chicago. My friend Bob Bean and I used to go to Howard Street a lot. There was really nothing else for us to do, for at that time we were both short and fat and covered with pimples and couldn't get dates to save our souls; and there were a lot of bars on Howard

Street that would serve you beer even if you weren't quite twenty-one. Or even not quite nineteen, for that matter.

One night we were sitting at a bar and Bob said, "Put your head down for a minute." I did, and then he said, "How many cash registers are there behind the bar?"

"One," I said.

"Three," he said. "Keep your head down. Now how many people besides us are in this bar?"

"Twelve?" I said.

"Eight," he said.

And that started us on a game that we played, off and on, for more than three years.

We'd walk into a bar, order a beer, and spend exactly ten minutes looking around, studying and memorizing every detail we could. After ten minutes we'd each put our heads down and start asking each other questions.

"How many chairs are there in here?" "How many windows?" "How many steps from the door to the bar?" "What color are the bartender's eyes?" "What's the ceiling like?"

After a couple of months we got so good at it, it was hard for either of us to ask a question the other couldn't answer.

"How many bottles are there behind the bar?" "Describe every picture and sign on the walls." "What was rung up on each cash register when we walked in?"

By the time we stopped playing, there was nothing we didn't know.

"Tell me the name of every bottle that's behind the bar." "Now, how full is each bottle? Half full? A quarter full? Three-quarters full?"

Really, we could do it.

"How many slats are there in those blinds that cover the front window?" "Describe in detail every person here." "How many glasses and bottles are there on each table?"

We had discovered the magic of seeing.

Years later I was working with another friend, Hal Silverman. Hal is an artist, one of those irritating people who can draw what he sees. He was drawing a chair. "Boy," I said, "that looks great—just like a chair. I wish I could do that."

"Do what?" he said.

"Draw something that looks like what it is."

"Why can't you?"

"I don't know; I just can't. If I tried to draw that chair it'd probably come out looking something like a chicken."

"You got something the matter with you?"

"What d'you mean?" I said.

"Can you print numbers and the alphabet? Can you write your name?"

"Of course."

"You got Saint Vitus' dance or arthritis or dyslexia or something?"

"No."

"Your eyes are OK?"

"Sure."

"Then why can't you draw what you see?"

"I don't know; I just can't."

Hal shook his head. "If there's nothing physical that prevents you from drawing that chair, it must be something mental that prevents you from doing it."

"Huh?"

"You've got a good command of your motor functions, your eyes are fine, you're not in pain, so the reason you can't draw this chair must be because you're not seeing the chair."

"Of course I can see the chair."

"Agreed. You can, but you're not."

"What d'you mean, I'm not?"

"If you really saw it, you could draw it. Here," he said, picking up the chair and handing it to me, "look at it for ten minutes. Study it. Take it apart in your mind. Then put it back together. Study its design, its shape, its form, its size, its materials, its construction, its colors. Look at how each piece of wood joins with every other piece. Notice how these pieces here are curved inward and these are curved outward. Concentrate. Take mental notes. Note that the back is longer than the legs, that the seat is wider than the back, that the front of the seat is wider than the back of the seat, that the legs splay outward slightly, that the back rest is bent backward. Count the cross members, notice how the turning on the legs is different from the turning on the arms. Look at it upside down, sideways, backward. Look at it. Work at it. If you do, you'll probably learn more about that chair in the next ten minutes than you have about any chair in your lifetime. And when you're through, you'll be able to draw something that actually looks like the thing you've learned about."

I did what he said. And he was right. After ten minutes, I was able to draw something that looked like a chair. Granted, the legs had a decided chickenlike quality to them, but still—it actually looked like a chair.

It's hardly necessary to point out that if you see things the way Bob Bean and Hal Silverman do, you'll be able to remember more of what you see.

I'm not saying that you'll remember everything you see. Nobody can do that. Nor am I saying you'll ever be as good as Hal Silverman is at drawing things. Some people are simply better at those kinds of things than other people.

But I am saying that by working at it you can see more and remember more of what you see than you ever dreamed you could. You can remember more about the people you meet and the places you go and the things you read.

And the more things you remember, the more things you'll have to combine in order to form new ideas.

But you must work at it. Every day.

Here's how to start:

Tomorrow morning on your way to work, or on your first coffee break, buy yourself a notebook. Not a loose-leaf notebook. Buy a ledger—something with a sense of permanence to it. Then every day write in it something that you've seen. Every day. It doesn't make any difference what you see; only that you see something and record it. (If you also want to write what you think about what you see, feel free. After all, that's what Thomas Wolfe and hundreds of other writers did and do.)

When your ledger is full, sit down and read it. Then start filling up another one. And another one. And another one.

For the rest of your life.

7.

Screw Up Your Courage

Honest criticism is hard to take, particularly from a relative, a friend, an acquaintance, or a stranger.

Franklin P. Jones

I'm not afraid to die. I just don't want to be there when it happens.

Woody Allen

No call alligator "long mouth" till you pass him.

Jamaican proverb

As I said, courage and curiosity are the two character traits all creative people seem to have.

But why do some people have these traits and others don't? And what can you do about it if you don't?

In the previous chapter we talked about curiosity and how to do deliberately what people who are curious do naturally.

But how can we become more courageous?

"An idea is delicate," said Charles Brower, the head of an advertising agency. "It can be killed by a sneer or a yawn; it can be stabbed to death by a quip and worried to death by a frown on the right man's brow."

I think this is why many people seem bereft of ideas.

They've run into too many sneers and yawns, they've heard too many quips. And so they've said the heck with it and don't even try to come up with ideas any more.

The fear of rejection shuts down their idea factories.

I can't tell you how to get enough courage to push ahead, to ignore the doubts and raised eyebrows and pooh-poohing you get when you tell people what you've got in mind.

All I know is that you must.

It will help to remember five things:

1. Everybody's Afraid—Everybody

The more naturally creative you are the more fear you probably feel, for your antennae are more finely attuned and you're more aware of what other people are thinking, more sensitive to their feelings, more affected by their actions. So it's only natural that you should get uptight and feel antsy and be afraid.

In the face of such fear it takes courage to speak out.

For courage—as Søren Kierkegaard and Ernest Hemingway and Friedrich Nietzsche and Jean-Paul Sartre and Albert Camus and others point out—is not the absence of fear. It is going ahead in spite of the danger, in spite of being afraid or feeling despair.

"Creativity is dangerous," wrote Robert Grudin in *The Grace of Great Things*.

"We cannot open ourselves to new insights without endangering the security of our prior assumptions. We cannot propose new ideas without risking disapproval and rejection."

Just remember, though, that the people who sneer or quip are afraid too. Afraid of your ideas.

That is often why they sneer or quip.

After all, ideas by their very nature are potentially destructive. They can change things. And the more original the ideas, the more radical the changes. And the more changes ideas wreak, the more they threaten people; the more they make people question their beliefs and actions, the more they make people anxious about their jobs and their futures.

So next time fight through your fear and blurt out your idea. If for no other reason than to make the other guy afraid.

2. There Are No Bad Ideas

Madame Curie had a "bad" idea that turned out to be radium.

Richard Drew had a "bad" idea that turned out to be Scotch tape.

Joseph Priestley invented carbonated water while he was investigating the chemistry of the air.

Blaise Pascal invented roulette while he was experimenting with perpetual motion.

Alexander Graham Bell was trying to invent a hearing aid when he invented the telephone.

Vulcanized rubber was discovered by accident by Charles Goodyear. So was antiknock gasoline by Charles Kettering. So was electric current by Luigi Galvani. So were potato chips by George Crum at the Moon Lake Lodge resort in Saratoga Springs. So was immunology by Louis Pasteur. So were X-rays by Wilhelm Roentgen. So was the telescope by Hans Lippershey. So was practical photography by Louis-Jacques-Mandé Daguerre. So was radioactivity by Henri Becquerel. So were friction matches by John Walker. So was the Slinky toy by Richard James. So was the microwave oven by Percy LeBaron Spencer. So was the pacemaker by Wilson Greatbatch. So was Teflon by Roy Plunkett. So was Krazy Glue by Harry Coover. So was Scotchgard by Patsy Sherman. So was penicillin by Alexander Fleming.

So was America by Christopher Columbus.

The moral? Never cry over spilled milk. Find a use for it. Or invent a better milk carton.

3. You Can Always Get Another Idea— Probably Even a Better One

In advertising, your ideas are always being turned down. It's the nature of the business.

And when they are turned down you grouse about it. You bitch and complain and swear and make idle threats and have too much to drink at lunch and go home early and yell at your kids.

Ralph Price had a different reaction, one that I've tried to acquire over the years.

"Rats," I'd say while leaving the client's building. "That was a great campaign they just turned down."

"Wow, this is super!" Ralph would say. "Now we can do a really great campaign."

You see, Ralph not only knew there was always another idea, he knew there was always a better idea.

If there wasn't, what were we doing in the advertising business? It was our job to come up with ideas. And if we couldn't beat our last idea—if our last idea was the best we could ever come up with—then we might as well quit, for we were on our way down, and pretty soon our bosses would get somebody who could come up with a better idea.

So Ralph never looked at a rejection as a defeat. It was an opportunity to do something better.

But even if you can't emulate Ralph when your idea somehow doesn't work, remember that at least you've found out what doesn't work, and that should help you get an idea that does.

It certainly helped Thomas Edison. In attempting to make a light bulb, he tried over a thousand ideas before he hit the one that worked.

Ray Bradbury wrote at least one short story every week for ten years before he wrote one that made the hair on his neck stand up.

Orville and Wilbur Wright worked for years trying to build an airplane. For the wings alone, they tested more than 200 designs in a wind tunnel they built.

Johannes Kepler spent nine years and filled 9,000 folio sheets with calculations in his small handwriting trying to work out the orbit of Mars before he concluded that the paths of the planets were not circular but elliptical.

So don't think that your idea is the end of the line. It's the beginning of another line.

4. Nobody Is Ever Criticized for Getting Too Many Ideas

Perhaps one of the things that is inhibiting you is the fear that your reputation, your future even, is riding on the idea you are about to suggest.

Perhaps it is, perhaps the sky will fall, perhaps people will laugh at you, or perhaps your idea won't work and will ruin the company you work for and you'll get fired and your family will disown you and your dog will run away and you'll die a pauper and a failure.

OK then, don't place all your dreams on one idea. Come up with a lot of ideas. That way you'll be known as "that genius with all the ideas" instead of "that jerk with the lousy idea."

5. Getting an Idea Is Worth It

It's a great feeling—swinging for the fences and connecting.

There's nothing quite like it. You're sitting in a room trying to come up with an idea, a solution, a way to go, and nothing is happening, and there is nothing there but walls and barriers and closed doors and stop signs and dead ends, and you're frustrated and worried and wondering if you'll ever find a way out of this maze, this box, this trap, when all of a sudden it hits, and wham—you see the whole thing, all at once, solved, with everything fitting and working together. Wheee.

"Creative achievement is the boldest initiative of the mind," said Robert Grudin, "an adventure that takes its hero simultaneously to the rim of knowledge and the limits of propriety.

"Its pleasure is not the comfort of the safe harbor, but the thrill of the reaching sail."

Swing for the fences. Go for broke.

Compared to a reaching sail, a safe harbor is pap.

8.

Team Up with Energy

Due to rising energy costs, the light at
the end of the tunnel has been turned off.

Unknown

I don't want any yes-men around me. I want
everybody to tell me the truth even if it costs
them their jobs.

Samuel Goldwyn

Teamwork is essential—it allows you to
blame someone else.

Unknown

We're all in this alone.

Lily Tomlin

Painting and writing and composing and sculpting—indeed, physically creating most forms of art—are, like brushing your teeth, jobs best done alone.

But when you're trying to get an idea, it often helps to do it with a friend.

Not simply a coworker or an acquaintance. A friend.

Or with a couple of friends.

The different experiences and frames of reference and points of view and backgrounds and needs and bits of knowledge other people bring to the effort often open doors to rooms you might not otherwise have known about.

And sitting inside one of those rooms, smoking a big fat cigar and looking swell, might well be an idea that solves your problem.

For make no mistake:

There is strength and spice and adventure and excitement and vitality and life and newness and power and energy in variety; in sameness there is lethargy.

This is the reason the people in advertising agencies who are charged with coming up with ideas—with thinking up the new products and ads and slogans and commercials and positionings—almost always work in twos or threes. They've learned from experience.

Indeed, of all the advertising ideas I've had over the years, I can remember only a few I can honestly call mine and mine alone.

Mostly my partner and I would be trying to come up with an idea for, say, a new soap for kids, and the conversation would probably go like one I remember having with two of my friends, Hal Silverman and Cliff Einstein. It went something like this:

Hal: Washing's a drag for kids. It'd be great if we could invent a soap that'd be fun to use.

Me: Why don't we call it Gorilla Soap? Or Giraffe Soap. You know how kids like animals. It'd be a kick.

Cliff: *Call* it gorilla, hell. Let's make it in the *shape* of a gorilla.

Me: Packaging would be a bear. No pun intended. Besides, after the first bath, the bar wouldn't look like anything, certainly not like a gorilla. All the fun'd be gone.

Hal: OK let's put the gorilla *inside* the soap, a prize, you know like Cracker Jack does, so when the kid finally uses up the soap, he gets a gorilla prize. Moms'll love it, too. It'd give the kid an incentive to wash, to get the prize.

Cliff: Each bar could have a different prize. A secret one. A gorilla in one, a car in another, dinosaurs, dolls for the girls, a secret decoder ring . . .

Me: So it's not Gorilla Soap anymore. It's Cracker Jack Soap.

Cliff: Better yet, we'll call it SOAPRIZE. (Hal and Cliff sold the idea to Dial Soap, which marketed it unsuccessfully.)

And this is the way Bill Bartley, an advertising agency art director, and I came up with an outdoor billboard for Knudsen yogurt in nine seconds flat:

Me: The client wants us to feature more than one flavor.

Bill: Oh, whoopee, hooray.

Me: Hooray for the red, white, and blue.

Bill: Hooray for the red, white, and blueberry.

And there we had it. The board had Bill's line and showed three cartons of yogurt—strawberry, vanilla, and blueberry.

Now I ask you: Whose soap idea and name was that, mine or Hal's or Cliff's? Whose yogurt idea?

More important: Would one of us have come up with those ideas working alone? Perhaps.

But even if one of us would have come up with them, I'm convinced the ideas surfaced faster with a couple of us working on them than they would have with just one of us working.

Let me tell you another story:

Years ago the Schick Electric Company came to our advertising agency with an assignment. "Here's something we made a couple of years ago," the company president said, "and we can't sell them. Got a warehouse full of the damn things."

The damn thing was an electrical device that heated cans of shaving cream. All the research showed that men thought it was a great idea. After all, who likes to lather up with cold shaving cream in the morning? But, as the president said, few bought it.

So we looked at the commercials they used to sell the device, wherein men used the product and said how it made shaving easier, and how nice it was to feel warm in the morning.

We liked the commercials. But Jean Craig, one of our copywriters, said: "I know what's wrong with that advertising. The commercials are beautifully done, but they're positioning the product all wrong. It's not something you should be trying to sell to men. It's something you should sell to women as a *gift* for men. Every Christmas I'm running around like crazy trying to find something to buy for my husband. This is perfect."

We did some focus groups with women. They, like Jean, were crazy about the product as a gift.

So we dubbed it "The Great Gift Invention of [that year]" and did a commercial showing guys waiting in line at a patent office with all sorts of inventions, including ours, which all the other inventors thought was a "great gift invention."

We put it on the air before Christmas in Chicago and Philadelphia and some other test markets.

Bingo.

The stores sold out of the dispensers the first week. We had to cancel the advertising and promise the stores we'd make more of the damn things for next year. Which the company did.

The next year we changed the copy in the commercial to "The Great Gift Invention of [that year]," went into more markets, and the same thing happened. And the next year, too.

By that third year it was delivering more profit to the company than any other product they had.

All because one of the people we teamed up with was a positive, energetic woman who had trouble finding a suitable Christmas gift for her husband.

So next time you're stuck with a problem, ask a couple of upbeat friends you get along with to help you kick around some ideas.

You'll be amazed, I think, at how smart and insightful and creative they are, at the different things they think of, the different roads they go down, the different doors they open.

Always remember three things, however:

1. Too many friends may rot the party.

Oh, I know that "brainstorming"—the classic idea-generating technique invented by Alex Osborne, himself an advertising man—calls for ten to twelve people, plus a group leader. And I know that thousands have used that technique successfully.

But—perhaps because of the peculiarities of the people I've worked with or the way we conducted the brainstorming sessions—in my experience in the workplace, that's way too many people.

Granted: it seems to work OK in the classroom. But two seem to work best at work; occasionally, three or four, no more.

2. Sometimes it's hard to make an ass out of yourself—to say the wild, far-out, stupid, unworkable idea that might trigger a great idea—in front of your boss.

So if you and your friend(s) are employed by the same company, you should be at the same level there. Unless, of course, one of your friends is a boss who's willing to make an ass out of herself.

3. As I mentioned in chapter 1, the group that has the most fun often has the best ideas. And since nothing "lethargizes" a group faster than a negative, sluggish member, make sure the friend(s) you team up with is (are) positive and energetic.

Follow David Ogilvy's advice: "Get rid of sad dogs who spread gloom."

9.

Rethink Your Thinking

Many people would rather die than think. In fact they do.

Bertrand Russell

Sixty minutes of thinking of any kind is bound to lead to confusion and unhappiness.

James Thurber

A conclusion is the place where you got tired thinking.

Martin H. Fischer

The way you think affects what you think about and what kinds of thoughts you get.

And the more kinds of thoughts you get, the more grist you'll have for your idea mill.

Here are some different ways of thinking:

Think Visually

You and I were brought up to think with words. And when we form a thought today—any thought—it's probably in the form of a statement. "Haste makes waste." "The world is all screwed up." "Nothing builds confidence like success."

But many of the most creative minds in history thought with pictures instead of words.

Albert Einstein said that he rarely thought in words. Notions came to him in images that only later he tried to express in words or formulas.

William Harvey was watching the exposed heart of a living fish when he suddenly "saw" it as a pump.

Frank Lloyd Wright thought of houses and buildings not as separate structures but as integral parts of the landscape.

Alfred Wegener noticed that the west coast of Africa fit into the east coast of South America and saw instantly that all continents were once part of a single continent.

Man Ray saw a woman's torso as a cello.

Albert Einstein wondered what the world would look like to a person riding a ray of light as it sped through space.

In trying to come to grips with the concept of infinity, David Hilbert, the mathematician, imagined a hotel with an infinite number of rooms that were all occupied. He then imagined a new guest arriving who asked for a room. "But of course," said the innkeeper, and he moved the person in room one to room two, the person in room two to room three, the person in room three to room four, and so on ad infinitum, thus freeing up room one for the new guest.

Lord Kelvin hit upon the idea of the mirror galvanometer when he noticed a reflection of light on his monocle.

Sigmund Freud conceived the idea of the sublimation of instinct by looking at a two-part cartoon— in the first picture, a little girl was herding a flock of goslings with a stick; in the second, she had grown into a governess herding a flock of ladies with a parasol.

Niels Bohr imagined in his mind's eye that an atom looked like our solar system.

Sir Isaac Newton saw in a flash that the moon behaved like an apple—that it "falls" just as an apple does.

And many of the creative people I worked with also think with pictures instead of with words.

If their company seems to be attracting new customers but is still losing sales, a picture of a leaking bucket springs to their minds.

If they have to do an advertisement on, say Master locks, they don't think of the lock as a lock, they think of it as a security guard, or as a watch dog, or as an insurance policy for your house or car or jewelry, or as a bodyguard for your children, or as something indestructible like the Rock of Gibraltar.

When Bill Bartley was charged with coming up with ads extolling a company's leadership position, he saw pictures of Winston Churchill giving his famous V sign, Robert E. Lee in battle, and Vince Lombardi being carried off the field by his players.

These people don't think words; they think pictures, they think relationships, they think metaphors, they think ideas.

"Once you get a visual idea," one of them said to me, "the words are easy." And he's right.

Once you think of a lock as a guard or a dog, it's easy to write headlines like: "Master lock. A security guard who never sleeps." Or "Now there's a security guard who never takes a day off." Or "Master locks. Think of them as watchdogs you don't have to feed." Or "We call our Master lock 'Fido.'" Or "We call our dog 'Masterlock.'"

Once you see your sagging sales as a broken toy or a drowning man or an out-of-date menu, the sooner you can begin fixing the toy or throwing out a lifeline or changing your offerings.

Once you visualize the problem of trying to get more shelf space for your grocery store product as one of those circus cars jammed full of clowns, or as a bathtub full of water, or as a suitcase with too many suits in it to close, the closer you are to finding a solution to the problem of how to get more shelf space when there is seemingly no more shelf space available.

Once you see the slowdown in your production line or distribution system as a neck in a bottle or a dam in a river or a stalled car on a freeway, the sooner you can start widening the bottle or bypassing the dam or removing the blockage.

So next time you're faced with a problem, try visualizing it instead of verbalizing it.

What does the problem look like? What does it resemble? What picture does it conjure up?

Think Laterally

You and I were also brought up to think linearly or vertically, to think logically from one point to the next until we reach a sound conclusion, to place one brick on top of another.

Such thinking is analytical, sequential, purposeful. If something doesn't make sense as we move along, we stop and go in another direction, taking one logical step after another until we reach a sound conclusion.

But there is another way of thinking, popularized by Edward de Bono, called lateral thinking.

In lateral thinking you make jumps. You don't have to follow the logical path; you can take side trips down alternate roads that seemingly don't lead anywhere.

Granted, it's impossible to tell after the fact how a problem was solved, laterally or vertically. That's because all good solutions make sense and thus have logical pathways to them.

But even though most solutions are obvious in hindsight, it is difficult to imagine some solutions being arrived at logically.

Let me give you an example:

A small company was having trouble with tardiness. Every week all twenty employees seemed to be getting in later and later.

The owner talked with people individually (a vertical solution). A little improvement but not much.

Then he called them all together and voiced his concern (another vertical solution). A bit more improvement. But a month later it was just as bad if not worse than before.

Then he did something that cured the problem and kept it cured.

He took Polaroid pictures of the office every fifteen minutes beginning at 9:00 in the morning when the office opened. At noon he posted the pictures on the bulletin board and marked the times on them.

The 9:00 picture showed nobody at work.

The 9:15 shot showed only one person in place.

The 9:30 picture showed eight people in place.

The 10:00 picture showed five people still missing.

The next day's pictures showed some improvement; the next day's, a dramatic improvement.

And by the end of the week everybody was there at 9:00. After another week he stopped the practice and hasn't had a problem since.

Could he have arrived at that solution through logical, vertical thinking? Probably.

Did he? I doubt it.

Let me give you another example:

There was a large telephone company in the Midwest, so the story goes, that right after the Korean War developed a great program for training supervisors of telephone operators.

Graduates of the program were absolutely top-notch and the program was awarded many commendations, written up in magazines, and studied by other organizations.

But there was a problem: The graduates were so good that as soon as a class graduated, most of them were contacted by other companies and hired away.

A car company might need three or four supervisors for their communications staff; an oil company, five or six; the government of Canada, ten. Even other telephone companies were raiding their sister company for supervisors.

The telephone company tried everything to keep its graduates—it gave them more money and fancy titles, it erected an "Honor Wall" where their names were inscribed, it furnished them expensive jackets to wear on the job, it sent their spouses flowers on every anniversary, it gave them extra vacation time—all to no avail.

The other companies simply gave them more money and even fancier titles and even more vacation time.

As you can imagine, the heads of the telephone company had many meetings trying to figure out ways to keep graduates from leaving. During one of these meetings, so the story goes, one of the managers lost his temper and shouted: "I'd like to chop their damn legs off—then they couldn't leave."

Everybody laughed. Except one person, who said, "Yes, of course, that's it."

"What's it?" said his boss.

"Why, we'll hire only handicapped people in wheelchairs for the program," he said. "We'll redo all our entrance ways, our elevators, our toilets. We'll furnish modified cars for them to drive to and from work. We'll work with doctors and physical therapists to develop exercise programs. We'll . . ."

And that's what they did.

And the other companies no longer tried to hire the graduates away because they knew they'd have to redo all their entrance ways and elevators and toilets, and modify cars, and so on.

And it all started because somebody suggested chopping "their damn legs off"—a very illogical solution.

That is lateral thinking.

Mr. de Bono wrote a number of books explaining the differences between lateral and vertical thinking and showing how to solve problems by thinking laterally. I recommend them to you.

Don't Assume Boundaries That Aren't There

If you're like most people, many times your thinking is inhibited because you unconsciously assume that a problem has restrictions and boundaries and limitations and constraints, when in fact it doesn't.

For example, if someone asks you to plant four trees so that each tree is exactly the same distance from each of the other trees, you'll probably automatically assume that the trees must be planted on a level piece of land. (I certainly did the first time someone gave me that problem.) And so you'll try to arrange four dots on a piece of paper so that each dot is the same distance from every other dot and quickly discover that you can't do it.

It isn't until you break that assumption about all the trees being on the same plane that you can solve the problem. Then you simply plant one tree at the top of a hill and plant the other three trees on the sides of the hill and bingo—your problem's solved.

But please note that you're the one who created the barrier to solving the problem, because you're the one who assumed that the trees had to be planted on a level piece of land.

Or take the famous nine dots, four lines problem. (You probably know this one, but no matter; it is the classic example of creating boundaries.) There are nine dots arranged like this:

•   •   •

•   •   •

•   •   •

Your job is to draw four straight lines through these dots without retracing or raising the pencil from the paper.

As long as you assume (as most people do) that the lines must not extend beyond the boundaries set by the outer line of dots, the problem is impossible to solve. But as soon as you allow your lines to travel outside those boundaries, a solution is possible.

Note that there was nothing in the problem as posed that said the lines must be kept within the dots. You unconsciously put that restriction upon yourself.

I used to line up my students against a wall in our classroom and ask them to make paper airplanes and toss them across the room to the opposite wall—about twenty feet away. They'd make all sorts of planes but they were never able to get one to go that distance.

Then I'd say, "OK you guys, now watch the world champion long-distance paper airplane builder in action." Whereupon I'd wad up a piece of notepaper into the size of a golf ball and pitch it underhand to the wall. Bingo.

Who said paper airplanes have to look like paper airplanes?

Here's another exercise I used to give to my students:

"Imagine," I'd say, "a piece of pipe about eighteen inches long and slightly larger in diameter than a Ping-Pong ball. One end of the pipe is welded tight to the floor.

"I'm going to drop a Ping-Pong ball into that pipe, give you the following things, and ask you to get the Ping-Pong ball out of the pipe.

"The things I'm going to give you are: last Sunday's newspaper, a pair of leather gloves, a book of matches, an eight-inch screwdriver, a fourteen-inch shoelace, four toothpicks, a package of chewing gum, and a straight-edge razor blade."

Over the years I got hundreds of ideas on how to get that Ping-Pong ball out of the pipe. Most of them would have worked; some were wonderfully imaginative.

Here are a few:

"Tie the shoelace around the handle of the screwdriver and drop the screwdriver into the pipe. Tear the newspaper into long strips. Stuff the strips into the pipe, jamming them down with enough force to cause the screwdriver to puncture the Ping-Pong ball. Remove the strips. Carefully lift up the impaled Ping-Pong ball."

"Lengthen the string by tying one end of it around the middle finger of the glove. Chew the pack of gum. Encase the other end of the string in the gum. Lower the wet gum down to the ball. When the gum has dried enough to adhere to the ball, raise the string."

"Cut the fingers off the gloves with the razor blade. Fill the fingers with pieces of newspaper. Light the finger-filled newspapers with the matches and hold them, one after the other, over the pipe. The fire will draw the oxygen out of the pipe, thereby lifting the Ping-Pong ball. When the ball gets close to the top, stab it with a toothpick."

"Tie one end of the shoelace around the end of a match, the other around the end of the screwdriver. Light the match. Quickly lower it into the pipe. When the lit match touches the Ping-Pong ball, it will adhere to it. Allow the match to cool. Raise up the now united ball and match."

It was amazing how many great ideas I got, proving once again that there are many, many ways to solve problems.

But nobody ever suggested pouring water into the pipe.

The reason of course is because my students (and you too probably) limited themselves to solving the problem with the things I gave them. But I never told them they had to use those things and only those things to get the Ping-Pong ball out of the pipe. They put that limitation on themselves.

Next time you have problems solving a problem ask yourself: "What assumptions am I making that I don't have to make?" "What unnecessary limitations am I putting on myself?"

Set Some Limits

"Wait a minute," I can hear you saying. "Didn't you just tell me not to put any unnecessary limitations on myself? Now you're telling me that I should set some limits. What's going on here?"

The limitations I talked about earlier were the imagined boundaries, the unconscious assumptions we often make about the nature of the problem.

Now I'm talking about the need to have a framework within which to work at finding a solution.

I know this sounds like a paradox—creativity needing a framework. "Are you crazy?" I can hear you saying again. "The creative mind should be free to roam, to explore, to seek wherever it wants. Put limits on it and it will shrivel up like a worm in the sun."

Agreed. It is a paradox. In *The Courage to Create*, Rollo May calls it a "phenomenon." But he explained "that creativity itself requires limits, for the creative act arises out of the struggle of human beings with and against that which limits them."

Let me give you an example:

When giving a team an assignment to create, say a television commercial, I found that if I gave them complete freedom they floundered. Too much freedom leads to chaos. But when they were forced to work within the guidelines of the creative strategy (see chapter 10) and a budget and a thirty-second length and an established theme line and of course a deadline, they always came up with solutions.

Joseph Heller found the same thing: "The ideas come to me; I don't produce them at will. They come to me in the course of a sort of controlled daydream, a directed reverie. It may have something to do with the disciplines of writing advertising copy (which I did for a number of years) where the limitations involved provide a considerable spur to the imagination."

"Small rooms discipline the mind; large rooms distract it," said Leonardo da Vinci.

"There's an essay of T. S. Eliot's," continued Heller, "in which he praises the disciplines of writing, claiming that if one is forced to write within a certain framework the imagination is taxed to its utmost and will produce its richest ideas. Given total freedom, however, the chances are good that the work will sprawl."

Duke Ellington composed music within the limits of the instruments he was writing for and the players who would play those instruments. "It's good to have limits," he said.

Walter Hunt was being dunned for money. He decided to invent something that was sorely needed, something so simple he could make a sketch of it in a few hours. (Talk about limits.) He invented the safety pin.

The Caesar salad was invented because the chef was forced to make something out of the ingredients he had. So was chicken marengo. So was bread pudding. And so, probably, was boiled lobster.

John Dryden said he preferred to write verse that is rhymed because "I often had a very happy thought as a result of looking for a rhyme."

Rollo May agreed: "When you write a poem, you discover that the very necessity of fitting your meaning into such and such a form requires you to search in your imagination for new meanings. You reject certain ways of saying it; you select others, always trying to form the poem again. In your forming, you arrive at new and more profound meanings than you had ever dreamed of."

The most stimulating limitation I've ever found is time. Deadlines spur you to get something accomplished.

Give yourself one.

10.

Learn How to Combine

He can beat your'n with his'n and he can beat his'n with your'n.

Football coach Bum Phillips on
the ability of coach Don Shula

Asthma doesn't seem to bother me any more unless I'm around cigars or dogs. The thing that would bother me most would be a dog smoking a cigar.

Steve Allen

Dr. Livingston I Presume (full name of Dr. Presume).

Unknown

To be is to do.   *Jean-Jacques Rousseau*
To do is to be.   *Jean-Paul Sartre*
Do be do be do.   *Frank Sinatra*

If "a new idea is nothing more nor less than a new combination of old elements," it stands to reason that the person who knows how to combine old elements is more likely to come up with a new idea than a person who doesn't know how to combine old elements.

Here are some suggestions that will help you combine:

Look for Analogues

An analogue is a comparison between two things that are similar in one or more respects, and is used to help make one of those things clearer or easier to understand.

Is your problem similar to other problems? What's it dissimilar to?

If the greatest benefit of your product or service is speed, what's the fastest thing in the world? Can you compare your benefit to that thing? What's the slowest thing in the world? Can you compare it to that?

If its greatest benefit is strength, what are the strongest and weakest things you can think of? Can you compare it to them?

Or what if it's convenience? Or economy? Or dependability? Or simplicity? Or durability? Or whatever? What are the most convenient, economical, dependable, simple, durable, or whatever things or people or ideas you can think of? Or the most inconvenient, uneconomical, undependable, complicated, fragile, or un-whatever things or people or ideas you can think of?

Literature is replete with analogues. A poem about writing a poem, or a story about building a house, or an essay about flying a kite could all be analogues for living a life. A novel about a baseball game could be an analogue for good and evil. A story about cheating at cards could be one for adultery.

Whatever problem or story line or invention or project you're working on, see if you can compare it, at least in one particular, to something else. It could help you arrive at solutions you might not have arrived at otherwise.

Break the Rules

Every activity has its rules and conventions and ways of doing things. They may not be etched in stone but they are etched in people's minds. Depend on it.

Most of the great advances in the sciences and arts—indeed, in everything—are the result of somebody breaking those rules.

Vincent van Gogh broke the rules on what a flower should look like.

Pablo Picasso broke the rules on what a woman's face should look like.

Sigmund Freud broke the rules on how to treat illness.

Louis Pasteur broke the rules on how to treat diseases.

Nicolay Ivanovich Lobachevsky broke the rules of Euclidean geometry.

Dick Fosbury broke the rules on how to high jump.

Pete Gogolak broke the rules on how to kick a football.

Igor Stravinsky broke the rules on how ballet music should sound.

Ludwig van Beethoven broke the rules on how a symphony should sound.

David Ogilvy broke the rules on how copy writing should sound.

Gerard Manley Hopkins broke the rules on what poems should sound like.

e. e. cummings broke the rules on what poems should look like.

Charles Eames broke the rules on what a chair should look like.

Eero Saarinen broke the rules on what a table should look like.

Antoni Gaudí broke the rules on what a building should look like.

Henry Ford broke the rules on how much workers should be paid.

Antonin Carême broke the rules on what a dessert should be.

Fannie Farmer broke the rules on what a cookbook should be.

We could go on all day with this; maybe all week. Suffice to say, rules are a great way to get ideas.

All you have to do is break them.

Play "What If?"

"What if?" is the game many advertising agency creative people play when trying to come up with a different way to present the benefits of a product or service.

What if we turned the product or service into a person, what kind of a person would it be? A man? A woman? A truck driver? An artist? A basketball player? What would that person say? How would that person act? What would that person sound like?

What if we turned it into an animal, what kind would it be?

What if we made the product smaller? Or larger? Or a different shape? Or a different color? Or lighter? Or heavier? Or packaged it differently? Or made it twice as strong? Or half as reliable? Or twice as reliable?

What if we made the service faster? Or cheaper? Or more convenient? Or more friendly? Or less friendly? Or more efficient?

Or slower? Or more expensive? Or less convenient? Or less efficient?

If we could add anything we wanted to the product or service, what would we add?

If we could subtract anything we wanted from it, what would we subtract?

What if it were suddenly invented or discovered today, for the first time, how would we introduce it?

What if a woman from Mars saw this product or service? How would you describe it to her? What would she think? Would she even want it?

What if the greatest benefit of this product or service were suddenly made illegal, what would you do? What if nobody wanted that benefit? What if everybody wanted that benefit?

What if we were able to make that benefit twice as powerful? Or half as strong? Or twice as important to people? Or half as important to people? Or more accessible? Or less accessible?

What if this product or service were the only one to provide this benefit? What if all of our competing products or services also offered it?

What if we went back in time to the 1800s, how would people react to this product or service? What if we went forward a couple of hundred years?

Play the same game when trying to solve a problem.

What if the problem were twice as bad as it is, then what would you do? How about ten times as bad? Or half as bad?

What if everybody had this problem?

What if nobody but you had it?

What if your biggest competitor had it?

What if we turned this problem upside down, what would it look like?

What would it look like backward?

What if this problem still exists next year, what will you do? How about ten years from now?

What if this problem suddenly didn't bother anybody anymore, what would you do?

If you had the opposite problem from the one you have now, how would you solve it?

What if someone in another field—such as the music business or the airline industry or the used car business—had this problem, how would they solve it?

How would an architect solve it? A plumber? A surgeon? A poet?

What if someone gave you a million dollars in cash to solve this problem, how would you spend the money?

Remembering that people are the cause of 99 percent of all problems, how would you solve it if you could fire anybody you wanted? Or hire anybody you wanted?

What if you were the problem, how would you change?

What if your best friend were the problem, what would you say?

What if you were a child, how would you solve the problem?

Look to Other Fields for Help

In *A Whack on the Side of the Head*, Dr. Roger von Oech wrote this insightful paragraph:

> I have consulted for the movie and television industries, the advertising industry, high technology research groups, marketing groups, artificial intelligence groups, and art departments. The one common denominator I have found is each culture feels that it is the most creative, and that its members have a special elixir for new ideas. I think this is nice; esprit de corps helps to create a good working environment. But I also feel that television could learn one heck of a lot from software people, and that R & D people could pick up a few ideas from advertising. Every culture, industry, discipline, department, and organization has its own way of dealing with problems, its own metaphors, models, and methodologies. But often the best ideas come from cutting across disciplinary boundaries and looking into other fields for new ideas and questions. Many significant advances in art, business, technology, and science have come about through the cross-fertilization of ideas. And to give a corollary, nothing will make a field stagnate more quickly than keeping out outside ideas.

The coin punch and the wine press were around and in constant use for centuries before Johannes Gutenberg saw the relationship between them and invented the printing press.

James J. Ritty was trying to figure out a way to record the cash taken in at his restaurant to dissuade his cashiers from pocketing so much of it. On a transatlantic steamer he saw a device that counted and recorded the turns of the propeller. He used the same principle to build the world's first cash register.

Charles Darwin credited a chance reading of Thomas Malthus's *Essay on Population* as the key that unlocked the mystery of evolution by natural selection.

Malthus showed that population was retarded by such "positive checks" as disease, accidents, war, and famine. Darwin wondered if similar circumstances might retard the growth of plants and animals, if their "struggle for existence" affected their fate.

"It at once struck me," he wrote, "that under these circumstances favorable species would tend to be preserved, and unfavorable ones to be destroyed. The result of this would be the formation of new species."

Benjamin Huntsman, a clock maker, was trying to improve the steel that watch springs were made of. He noticed that the ovens that local glassmakers used were fired with coke and lined with Stourbridge clay. He tried the same thing and "crucible steel" was born.

George Westinghouse got the idea for air brakes while reading about a compressed-air rock drill they used for tunneling in the Alps.

Jim Crocker, a NASA engineer, got the idea for how to fit correcting mirrors into the Hubble telescope from a European-style showerhead he saw in a German hotel.

Before René Descartes, there was no such thing as analytical geometry; arithmetic and geometry were separate.

So were the sciences of electricity and magnetism before Hans Christian Oersted, Owen Willans Richardson, Michael Faraday, and others created the field of electromagnetism.

So were astronomy and physics before Johannes Kepler borrowed from each to create modern astronomy.

Something is going on right now in some other field that could help you solve your problem, that could give you a fresh insight, that could turn your thinking in a new direction, that you could combine with something you already know, that you could use to unlock your mystery.

Keep your eye and ear out for it.

Take Chances

Getting an idea usually means combining things that were never combined before—in other words, taking chances. So by definition you must take a chance if you are to get an idea.

Never forget this, for if you don't take a chance, you won't get an idea.

"The only cats worth anything," said philosopher and pianist Thelonious Monk, "are the cats that take chances. Sometimes I play things I never heard myself."

Play something you never heard yourself.

Part II

A Five-Step Method

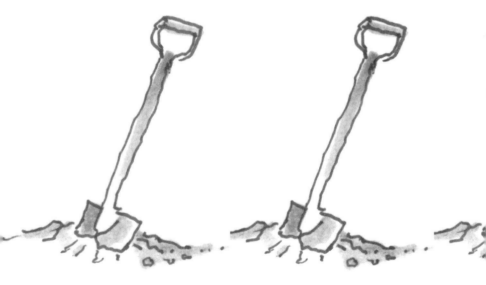

for Producing Ideas

11.

Define the Problem

No problem is so big or so complicated that it can't be run away from.

Charles Schultz

It is better to know some of the questions than all of the answers.

James Thurber

Computers are useless. They can only give you answers.

Picasso

There is no human problem which could not be solved if people would simply do as I advise.

Gore Vidal

$\mathcal{S}$ince all problems have solutions, it's critical that you define your problem correctly.

If you don't you might solve the wrong problem.

In advertising—the field that I'm familiar with—the statement of the problem is often called a creative work plan or a creative strategy or a mission statement or some such thing. It demands answers to questions like, "What are we trying to say and why are we trying to say it?" "Who are we trying to say it to and why?" "What can we say that our competition can't?" "What's our product's or service's reason for being?"

These plans are essential, for as Norm Brown, the head of an advertising agency, once said: "If you don't know where you're going, every road leads there."

Every field has its own kind of plan that sets forth objectives and missions and strategies—what the problems are, what the opportunities are, what needs to be done.

And "The formulation of a problem," wrote Albert Einstein, "is often more essential than its solution, which may be merely a matter of mathematical or experimental skills. To raise new questions, new problems, to regard old problems from a new angle, requires creative imagination and makes real advances."

He's right of course. For even such a simple question as "How can I do all this work on time?" is vastly different from "How can I get all this work done on time?"

The first question will result in all sorts of labor-saving techniques and shortcuts; the second, in dividing the work load up among others.

It is said that Henry Ford invented the assembly line simply by changing the question from "How do we get the people to the work?" to "How do we get the work to the people?"

Edward Jenner discovered the vaccine for smallpox simply by changing the question from "Why do people get smallpox?" to "Why don't milkmaids get smallpox?"

Grocers used to fetch the groceries for the customer. And they were always trying to improve their service by asking, "How can I get the groceries faster for my customer?" Then somebody invented the supermarket by asking, "How can the customer get the groceries for me?"

"The greatness of the philosophers of the scientific revolution," wrote Arthur Koestler, "consisted not so much in finding the right answers but in asking the right questions; in seeing a problem where nobody saw one before; in substituting a 'why' for a 'how.' "

Jonas Salk agreed: "The answer to any question 'pre-exists.' We need to ask the right question to reveal the answer."

So take care in what questions you ask, in how you define your problem.

If you're having trouble solving it or if your solutions seem flat somehow, try defining the problem differently and then solve it.

Let me give you a couple of examples:

Assume that you are the manager of a ten-story office building that was built back in the days when everybody had big, spacious offices. Back then two elevators were sufficient to handle the number of people working in the building. But over the years the large offices were converted to smaller offices, and now it's obvious that the building's two elevators cannot handle the number of workers.

You've installed the fastest, most efficient, and up-to-date computer-operated elevators made, yet every morning and every afternoon crowds of angry employees still gather in the lobbies grousing about having to wait for three minutes or more before they can catch a ride. Complaints rain down upon your head. Tenants are threatening to leave. It's crisis time.

What do you do?

If you think the problem through logically (or vertically, if you'll excuse the pun) it seems obvious that you have to figure out a way either (a) to get more people up and down the building faster, or (b) to reduce the number of people going up and down the building at the same time. You could, therefore:

Make the elevator shafts larger and put in larger elevators.

Or bore a hole through the building and install a couple more elevator shafts.

Or attach a couple of elevator shafts to the outside of the building.

Or turn the stairways into escalators.

Or attach escalators to the outside of the building.

Or connect every other floor with escalators, thus reducing by half the number of floors each elevator has to stop at.

Or give monthly prizes to the earliest-arriving and latest-leaving workers, thus reducing the number of workers using the elevators at the busiest times.

Or work with the various employers in the building to stagger their starting and quitting times.

Or assign staggered boarding times for each floor, thus limiting the number of people going up and down at any one time.

Or get the fire department to limit the number of people allowed in the building and/or the lobbies at any one time.

Or sponsor programs that extol the benefits of stair climbing.

All these ideas are good ones of course (albeit some are rather expensive), and all would probably work to one degree or another.

But when the manager of an office building in Chicago was faced with this identical problem, she did none of these things.

Instead, she installed wall-to-wall, floor-to-ceiling mirrors in every elevator lobby. She figured (correctly as it turned out) that people wouldn't mind waiting so much if they could spend that time looking at themselves.

In other words, she solved a different problem.

Instead of trying to figure out how to add elevators and escalators or how to reduce the number of people riding them, she changed the problem and asked herself, "How do I make waiting less frustrating?"

Or assume you're the police chief in the 1960s of a town by the ocean. The town is one of those meccas for vacationing college students during spring break.

The businesspeople in town love the money the students bring in, but every year the students (males mostly, for this is before women's lib) are getting more and more unruly.

Worse, putting them in jail overnight for being drunk and disorderly or for disturbing the peace or for behaving obscenely or for damaging property isn't helping. Indeed, it only seems to exacerbate the problem, for jail time is becoming a badge of honor, of respect, of machismo. If a student hasn't been in jail he isn't part of what's happening, he isn't in, he isn't a man.

So you decide to get tough: You put them on bread and water.

Wrong.

Now guys who don't even drink start feigning drunkenness just so they can be arrested, just so when they get out of jail the next day they can brag about being in jail on bread and water. Suddenly students who haven't been in jail are sissies.

You run out of jail space and have to bring in mobile jails from the next county. Your jail staff is working overtime. The problem is getting out of hand.

You're in a bind. You must enforce the law; that's your job. But when you enforce it you make the problem worse.

What do you do?

There are a number of things you could do; there always are. But when this actually happened to a police chief in Florida in 1963, this is what he did:

He put the jailed students on baby food.

Instead of treating them like criminals, he treated them like infants. And almost overnight he turned macho students into laughingstocks.

The police chief was a quick learner.

The first time he asked himself, "How do I more severely punish these students for breaking the law?" And he put them on bread and water.

When that didn't work he asked himself, "How do I embarrass these students for breaking the law?" And bingo.

Many times it's like that: You simply rephrase the problem and bingo—all sorts of different solutions appear.

Or pretend that you're in charge of burying the dead during the Black Plague.

You're under orders to bury each person in a coffin as quickly as possible in order to help slow down the spread of the disease. But in your rush to bury the dead you discover, just in the nick of time, that someone is still alive.

You're horrified. How can you make sure, you ask yourself, that you're not burying someone else alive?

You ask doctors for help. They talk about checking for heartbeats and signs of breathing, but the people who are carting the bodies away have neither the energy nor the desire to check every one. Too many people are dying too fast.

What do you do?

Legend has it that when a man in England faced this problem he simply changed the question from "How do I make sure I'm not burying someone alive?" to "How do I make sure that everybody I'm burying is dead?"

Bingo.

Now all he had to do was install three upright stakes in the bottom of every coffin. If the person was not dead before he was placed in the coffin, he surely was after.

Businesspeople ask the wrong questions all the time. Many times these questions are based on assumptions so deep-seated they don't even know they're making them.

Let me give you an example:

I once worked for a donut company. It operated hundreds of stores where they made and sold donuts.

Over the years their sales were gradually going down and they asked us to come up with some ideas on how to increase traffic; on how, in other words, to get more customers into the stores.

"Why not try to get your existing customers to buy more donuts?" we asked.

"Because our sales figures show that we're getting fewer customers every year, not that our customers are buying fewer donuts."

We discussed a number of ways we could attract more customers, including adding muffins and scones and sweet rolls to the menu, distributing coupons in the neighborhoods surrounding the stores, offering special prices at slow hours, offering free coffee with every order, coming up with new advertising, directing our advertising at teens, at women, at office workers, and so forth.

Then we made a suggestion that startled them: "If you want more customers, you might want to reconsider the question."

"How do you mean?"

"Well right now you're asking, 'How do I get more customers to come to us?' "

"Yes?"

"But if you asked, 'How do I get more customers period?' or simply, 'How do I sell more donuts?' your whole marketing approach might change."

"Say again?"

"If you asked either of those questions, you might eventually stop thinking of your stores as retail outlets and start thinking of them as individual manufacturing plants."

"What are you talking about?"

"If they were donut manufacturing plants, your stores would sell donuts retail just like they're doing now; but in addition they'd probably hire salesmen to go out into their marketing areas and drum up more business."

"From where?"

"From office buildings and schools, from apartment buildings, from convenience stores, from construction sites and factories and malls and condos, from gas stations, from wherever.

"Heck, they could even sell donuts to restaurants and coffee shops—those places have to get the donuts they sell from someone, don't they? Why not from you? And to bakeries—after all, most bakeries don't deep-fry dough, they bake it.

"Maybe you could even build some donut trucks, trucks that you could drive around selling hot coffee and donuts in the morning.

"And hire school kids to deliver donuts before school to the places that gave you standing orders.

"And you could even . . ."

But we had lost them. I think all they saw was the work and the risk involved, and so the idea was never tried.

But it shows, I believe, how a simple change in the question can revolutionize your thinking.

So if you're bogged down, try asking a different question.

If you've been asking "How can we make the production line more efficient?" try asking "How can we make the production line less inefficient?" Or "How can we change the production line so that the workers will enjoy their work more?"

If you've been asking "Why aren't people buying my product?" try asking "Why are people buying my product?" Or "Why aren't people who do buy my product buying it more often?" Or "Why aren't people who do buy my product buying more of it?" Or "Why are people buying my competitor's product?" Or "Why aren't people buying either of our products?" Or "How can I sell more of my product?" Or "What else can I sell that will help the sales of my product?" Or "Where else can I sell my product?" Or "What else can my product be used for?" Or "How else can my product help people?" Or "How can I change my product to make it more desirable?"

If you've been asking "How do I save more money?" try asking "How do I spend less money?" Or "How can I get more money?" Or "How can I get more with the money I do spend?" Or "How can I get things for free?" Or "How can I do without money?" Or "How can I do without those things that I spend money on?"

If you've been asking "How do I get the salespeople to make more calls?" try asking "How do I get my salespeople to make fewer but more qualified calls?" Or "How do I get my salespeople to convert more of the calls they do make?" Or "How do I get my salespeople to call on more prospects at the same time?" Or "How can I get the prospects to call on my salespeople?" Or "How do I make it unnecessary for my salespeople to call on customers?"

Different questions, different answers, different solutions.

12.

Gather the Information

Aristotle was famous for knowing everything.
He taught that the brain exists merely to cool
the blood and is not involved
in the process of thinking.
This is true only of certain
persons.

Will Cuppy

If there is another way to skin a cat,
I don't want to know about it.

Steve Kravitz

We don't know a millionth of one percent
about anything.

Thomas Edison

Let me tell you a story:

It was my first year in advertising. Our agency just got a new account—a meat packer. The owner wanted us to advertise his bacon. I remember my first copy chief, Bud Boyd, saying that he wanted to ask him "a few questions" before we started to work.

"What is bacon, exactly?"

"What kind of hogs?"

"Do some hogs produce better bacon than other hogs?"

"Why?"

"What kind of hogs does your competition use?"

"What are the hogs fed?"

"Why are they fed corn and whey and peanuts and slop?"

"Where do the corn and whey and peanuts and slop come from?"

"What kind of corn?"

"What kind of whey?"

"What kind of peanuts?"

"What kind of slop?"

"How much of each are they fed?"

"Why?"

"Are the hogs your competition uses fed the same things, the same way?"

"Can you find out?"

"Do the hogs you use win prizes at state fairs?"

"How many prizes?"

"Is that more prizes than the hogs your competition uses?"

"Can you find out?"

"Who are the people who raise your hogs?"

"Are they all just from that one state?"

"Why?"

"What kind of buildings do the hogs live in?"

"Is the temperature or humidity or lighting controlled?"

"How are they shipped to market?"

"How old are they when they're shipped to market?"

"How much do they weigh when they're shipped to market?"

"Does any of this differ from what your competitors do?"

"Is there anything about your hogs that is different from the hogs your competition uses?"

"Could you arrange for me to talk on the phone to a couple of the people who raise the hogs?"

"How is bacon made?"

"What is it cut with?"

"Why is it the thickness that it is?"

"Why is it the length that it is?"

"Why is it the width that it is?"

"What are its fat and moisture contents?"

"Why aren't they lower?"

"Why aren't they higher?"

"Is any of this different from your competition's bacon?"

"When may we visit your packinghouse and talk to some of your people?"

"Why do you cure bacon?"

"What do you cure it with?"

"How long do you cure it?"

"Why do you smoke it?"

"How do you smoke it?"

"What kind of wood do you use?"

"Why?"

"How long do you smoke it?"

"Is any of this different from your competition?"

"Why is bacon packaged the way it is?"

"How can you tell if the bacon's not fresh?"

"Why does old bacon burn twice as fast as fresh bacon?"

"What makes one kind of bacon better than another kind of bacon?"

"What is the ideal proportion of fat to meat in bacon?"

"Why?"

"What is your bacon's proportion of fat to meat?"

"What is your competition's?"

"Does your bacon look different from your competition's?"

"Have you done any taste tests on your bacon?"

"Is there anything about your bacon that you'd change if you could?"

"What is the best way to cook bacon?"

"Why is frying better than broiling?"

"Why should you start with a cold frying pan?"

"Why should you turn it frequently?"

"Why should you pour off the grease?"

"My mother used to blanch bacon before she fried it. Is that a good idea?"

"Why not?"

"Do you have any books on bacon that I might read?"

All morning long and all through lunch Bud asked questions. When lunch was over the client said he had meetings to attend. Bud asked if we could come back tomorrow.

"What for?" said our new client. "I've told you everything I know about bacon, believe me."

"I just wanted to ask a few questions," said Bud, "about the people who make and package and deliver and sell the bacon. And of course about the people who buy and cook and serve and eat it."

Obviously Bud believed in getting as much information as he could about a subject before he started to come up with ideas about it.

So do I. And so does everybody I know of who writes about getting ideas.

In advertising it's easy to get the information. You just ask the client for it.

But you have to ask. And ask. And ask.

Another Bud—Bud Robbins, the head of an advertising agency—told this story:

"Back in the sixties, I was hired by an ad agency to write copy on the Aeolian Piano Company account. My first assignment was for an ad to be placed in the *New York Times* for one of their grand pianos. The only background information I received was some previous ads and a few faded close-up shots . . . and, of course, the due date.

"I volunteered I couldn't even play a piano let alone write about why anyone would spend $5,000 for this piano when they could purchase a Baldwin or Steinway for the same amount."

After much arguing, ". . . reluctantly a tour of the Aeolian factory in upstate New York was arranged.

"The tour lasted two days and although the care and construction appeared meticulous, $5,000 still seemed to be a lot of money.

"Just before leaving, I was escorted into the show room by the National Sales Manager. In an elegant setting sat their piano alongside the comparably priced Steinway and Baldwin.

" 'They sure look alike,' I commented.

" 'They sure do. About the only real difference is the shipping weight—ours is heavier.'

" 'Heavier?' I asked. 'What makes ours heavier?'

" 'The Capo d'astro bar.'

" 'What's a Capo d'astro bar?'

" 'Here, I'll show you. Get down on your knees.'

"Once under the piano he pointed to a metallic bar fixed across the harp and bearing down on the highest octaves. 'It takes 50 years before the harp in the piano warps. That's when the Capo d'astro bar goes to work. It prevents that warping.'

"I left the National Sales Manager under his piano and dove under the Baldwin to find a Tinkertoy Capo d'astro bar at best. The same for the Steinway.

" 'You mean the Capo d'astro bar really doesn't go to work for 50 years?' I asked.

" 'Well, there's got to be some reason why the Met uses it,' he casually added.

"I froze. 'Are you telling me that the Metropolitan Opera House in New York City uses this piano?'

" 'Sure. And their Capo d'astro bar should be working by now.'

"Upstate New York looks nothing like the front of the Metropolitan Opera House where I met the legendary Carmen, Risë Stevens. She was now in charge of moving the Met to Lincoln Center.

"Ms. Stevens told me, 'About the only thing the Met is taking with them is their piano.'

"That quote was the headline of our first ad.

"The result created a six-year wait between order and delivery.

"My point is this. No matter what the account, I promise you, the Capo d'astro bar is there."

And it is.

In the problem you're working on right now, there is some fact, some overlooked relationship with something else, some nugget of information, that will help you solve its mystery, that will help you unlock the door to its solution.

So if it's not easy for you to get the information you need, don't skip this step. It is essential.

It is the "specific knowledge" that James Webb Young talked about: specific knowledge that you need to combine with "general knowledge about life and events."

"A creative man can't jump from nothing to a great idea," said Bill Bernbach, the head of an advertising agency. "He needs a springboard of information."

Know that the nugget you're looking for is there, and know that you will find it, just as you know that the idea it will help form exists, just as you know that you will find that idea.

Dig for it.

The easiest way is, of course, by surfing the Internet.

A dozen or so years ago, when I was writing the original version of *How to Get Ideas*, I spent months at the UCLA Research Library looking through books for information and quotes and ideas on ideas. In writing this second edition, I spent days (not months) at home (not at the library) Googling for additional information and ideas.

If you have not yet learned how to use search engines like Google or Yahoo or Ask, learn. Go to one of them, type in the subject you're working on, hit "Search," and you're on your way.

But don't stop there.

Read books. Read magazine articles. Read newspaper articles. Consult the encyclopedia. Become more like a child again—ask questions. Ask why. Ask why not. Visit the plant. Visit the warehouse. Talk with the workers. Talk with the suppliers. Work in the store. Ride with the salespeople. Seek out customers and talk with them. Seek out noncustomers and talk with them. Seek out your competitor's customers and talk with them. Read your competitor's annual report. Talk with the engineers. Talk with the designers. Work on the truck. Work in the field. Sample the product. Sample your competitor's product. Go to lectures. Go to the library. Go to bookstores. Ask your friends. Ask your kids. Ask your mother.

But perhaps most important, put your mind on it.

It's amazing what happens when you keep something in the forefront of your consciousness.

Remember someone (I think it was Linus) in *Peanuts* telling Charlie Brown not to think about his tongue? The result was that his tongue was all he was able to think about for the next three days.

It's true. Think about anything and you'll see it, you'll hear it, you'll sense it all around you. Next time you take a walk put your mind on front doors or roofing materials and you'll see more kinds of front doors and roofing materials than you ever saw before.

And if it's true about white horses and makes of cars and front doors and roofing materials, it's true about ideas.

I once saw a TV interview with Eric Hoffer, the longshoreman/philosopher, and he said the same thing.

The interviewer asked him how he researched subjects for his books, how he got the information grist for his intellectual mill (or—if you prefer—how he got specific knowledge about a problem).

I don't remember Mr. Hoffer's exact words, but basically he said that he thought about the subject hard and continuously, and that as a result of that effort the information about that subject came to him.

"What do you mean—it comes to you?"

Mr. Hoffer said that if he was thinking about maintenance, for example, and how and why different cultures maintain things differently, then it seemed that every book he selected from the library shelf had something to say about that subject, every newspaper article mentioned it, things he saw and heard related to it; in short, he didn't have to go looking for information on his subject because the information came to him.

Thomas Mann said the same thing: "If you are possessed by an idea you find it expressed everywhere, you even smell it."

So keep your mind on it; become possessed; ask ask ask; dig dig dig. Do everything you can to get the information before you get to work.

It is the springboard you need for your leap.

13.

Search for the Idea

If there is no wind, row.

Latin proverb

The biggest sin is sitting on your ass.

Florynce Kennedy

 Writing is easy. All you do is stare at a blank sheet of
paper until drops of blood form on your forehead.

Gene Fowler

Cliff Einstein, the head of an advertising agency, says: "The best way to get an idea is to get an idea."

He means that once you have an idea, the pressure is off to have an idea.

He also means that ideas have a way of snowballing, that the best way to get the whole process going is to prime it with an idea, any idea. It doesn't matter if the idea makes sense or solves the problem or is even germane, just as long as it's something new and different.

I know this sounds crazy, but try it sometime. It really works. Say: "Why don't we paint it green?" Or "What if . . ."

Hal Riney, another agency head, said: "Actually, I suppose the creative process is probably nothing more than trial and error, guided by facts, experience, and taste."

Linus Pauling said: "The best way to get a good idea is to get a lot of ideas."

He was saying the same thing my friend in Chicago who gave the assignment on Swiss Army knives was saying—getting many ideas is easier than getting the impossible "right" one.

He was also saying that many times ideas don't make it in the real world. So the best way to cover yourself "is to get a lot of ideas."

But note one thing: All of these people are saying, "Do something, for heaven's sake. Don't just sit there and wait for an idea to come to you. Go after it. Work at it. Search for it. Do it."

Here's one of the exercises I gave my students:

"In the next ten minutes, I want you to give me fifty uses for a $2'' \times 2'' \times 2''$ block of wood."

Over the years I've gotten everything from "wrap it as a gift and send it to my mother-in-law," to "cut it into sixty-four squares and glue them together to form a chessboard," to "throw it at the next teacher who asks me to give him fifty uses for a $2'' \times 2'' \times 2''$ block of wood."

One thing I used to notice was that the students' ideas would come hesitantly at first, then faster, and finally they were coming too fast to even record them with a key word on the blackboard.

Many problems are like that block-of-wood problem.

At first ideas seem as hard to find as crumbs on an oriental rug. Then they start coming in bunches. When they do, don't stop to analyze them; if you do you'll stop the flow, the rhythm, the magic. Write them down and go on to the next one.

Analysis is for later.

Here's another thing I asked my students:

"What's half of thirteen?"

Someone would say, "Six and a half," and I'd write it on the blackboard.

"OK, what else is half of thirteen?"

Hesitantly someone would say, "Six point five?"

"Exactly. What else is half of thirteen?"

And they'd all give me the blank look that cows give to passing cars.

"OK," I'd say, "I want you to remember what you're thinking and feeling right now—that I'm crazy, that there are no other answers, that half of thirteen is six and a half or six point five and that's it.

"Now, think about it, think: What else is half of thirteen?"

"One and three," someone would finally say with a smile. A breakthrough.

"Exactly. What else is half of thirteen?"

"Four. Thirteen has eight letters. Half of eight is four."

"Exactly. What else is half of thirteen?"

"Thir and teen." They were getting into it now. Having fun.

"Exactly. What else is half of thirteen?"

A student would come to the blackboard, write THIRTEEN on it, erase the lower half of it, point to what was left and triumphantly say, "That's half of thirteen."

"Exactly. What else is half of thirteen?"

Then the same student would rewrite THIRTEEN

on the board, erase the upper half, and say the same thing again. Wheee.

"Exactly. What else is half of thirteen?"

Another student would come to the blackboard and do what the previous student did only with the numerals 1 and 3 instead of the word THIRTEEN.

"Exactly. What else is half of thirteen?"

Another student would come to the blackboard and do what the previous students did only with the lowercase word thirteen.

"Exactly. What else is half of thirteen?"

"Eight. Thirteen in Roman numerals is XIII. The upper half of that is eight." Another breakthrough. They were rolling.

"Exactly. What else is half of thirteen?"

A student would write the lower half of XIII on the board.

"Exactly. What else is half of thirteen?"

"Eleven and two. The right and left half of the Roman numeral XIII."

"Exactly. What else is half of thirteen?"

"One-one and zero-one. In the binary system, thirteen is written one-one-zero-one. So half is one-one and zero-one. Also eleven and one."

"Exactly. What else is half of thirteen?"

Someone would write 1101 on the board and erase the upper half, then write it again and erase the lower half.

"Exactly. What else is half of thirteen?"

"Two. One and three is four. Half of four is two."

Still another breakthrough.

"Exactly. What else is half of thirteen?"

Someone would come to the board and write | | | | | | | and then erase half of the last |.

"Exactly. What else is half of thirteen?"

Someone would come to the board and write | | | | | | | | | | | | | and then erase the upper half, write it again and erase the lower half.

"Exactly. What else is half of thirteen?"

"Three. Thirteen is the six-letter word *treize* in French." Another breakthrough. They were into foreign languages now. "Also the letters *tre* and *ize* because each of those is half of *treize*. Also the upper half of . . ."

"OK stop!" I'd say. "Remember back when we started? How you thought there was only one answer? Well now you know: There's always another answer. You just have to search for it."

And you do.

You must force yourself to look at the problem, to search for the idea, to find the solution the way Hal Silverman forced me to look look look at his chair.

Think laterally. Think visually. Play "What if?" Look for analogues. Look for things to combine. Ask yourself what assumptions you're making, what rules you're following. Ask yourself how a six-year old would solve it. Screw up your courage and attack.

If you need extra motivation to find the idea, do what the illustrator of this book sometimes does—he pretends that the idea he's looking for contains a hundred dollar bill. "If you really want to find what you're looking for, you'll find it," he says. "You always want to find a hundred dollars."

But at some point you must stop looking for it, you must stop thinking about it.

Oh, I know that continual, unrelenting effort often produces dramatic results.

Andrew Wiles worked for seven years before he proved Pierre de Fermat's Last Theorem—a proof that eluded thousands of mathematicians for centuries.

Richard Gatling spent four years working on a machine gun before he succeeded.

Nikola Tesla, the inventor of (among other things) alternating current, regularly worked from ten in the morning straight through until five the next morning, seven days a week.

Thomas Edison's tenacity is a legend. So is Johannes Kepler's. And Albert Einstein's. And Sir Isaac Newton's. And Linus Pauling's. And and and and—the tenacity list goes on and on.

Still, there comes a time—and it will differ with every person and every problem—when enough is enough. You've (to paraphrase Koestler) uncovered, selected, reshuffled, combined, and synthesized all the already existing facts, ideas, faculties, and skills you can. And the idea still eludes you.

When that time comes follow the advice in the next chapter.

14.

Forget about It

It is sometimes expedient to forget who we are.

Publilius Syrus

Eric: My wife's got a terrible memory.
Ernie: Really?
Eric: Yes, she never forgets a thing.

Eric Morecambe and Ernie Wise

There are three things I always forget. Names, faces—the third I can't remember.

Italo Svevo

This is something you do only after you follow the advice in the previous chapter.

It is also something that I didn't get the chance to do often enough in advertising. Usually there wasn't time to forget about problems. You had to get ideas now. Not tomorrow. Now.

It's the same in journalism. Just listen to Andy Rooney: "The best creative ideas are the result of the same slow, selective, cognitive process that produces the sum of a column of figures. Anyone who waits for an idea to strike him has a long wait coming. If I have a deadline for a column or a television script, I sit down at the typewriter and damn well decide to have an idea. There's nothing magical about the process."

But I think Mr. Rooney is making a law out of a necessity.

I do not mean to disparage the hard work Mr. Rooney champions. As I pointed out in the previous chapter, it is essential.

But the weight of evidence suggests that when you're having trouble solving a problem or coming up with an idea, forgetting about it is also essential.

166

Just listen:

Hermann von Helmholtz said: "So far as I am concerned they [ideas] have never come to me when my mind was fatigued or when I was at my working table."

Albert Einstein said his best ideas came to him while he was shaving.

Grant Wood said: "All the really good ideas I ever had came to me while I was milking a cow."

Henri Poincaré tells of working hard to solve a math problem. He failed, so he went on a holiday. As he stepped on a bus, suddenly the answer came to him.

"I have found," Bertrand Russell wrote, "that if I have to write upon some rather difficult topic, the best plan is to think about it with very great intensity—the greatest intensity of which I am capable—for a few hours or days, and at the end of that time give orders, so to speak, that the work is to proceed underground. After some months I would return consciously to the topic and find that the work had been done."

C. G. Suits, the legendary chief of research at General Electric, said that all the discoveries in research laboratories came as hunches during a period of relaxation, following a period of intensive thinking and fact gathering.

Rollo May believes that inspiration comes from sources in the unconscious that are stimulated by conscious "hard work" and then liberated by the "rest" that follows it.

"Saturate yourself through and through with your subject . . . and wait" is Lloyd Morgan's advice.

Indeed, as Philip Goldberg pointed out in *The Babinski Reflex*, this phenomenon (which he dubs the "Eureka Effect" after Archimedes and his bathtub discovery) occurs so often it "has been identified as a common feature of scientific discovery, artistic creation, problem solving, and decision making."

So when you get stuck on an idea or a project or a problem, or when the little ideas stop coming as fast as they did and you still don't have the big idea, or when it feels as if you're pounding your head against an iron gate, or when things get labored and difficult, or whenever that little voice inside you starts saying, "This isn't working," forget about it and work on something else.

Note that I didn't say forget about it and relax, or forget about it and vegetate, or forget about it and watch sitcoms on TV for a week.

I said forget about it and work on something else.

In my experience, mental relaxation (except for meditation) is overrated. It might even be counterproductive, for it stops momentum, it

suffocates your interests, it shuts down the effort
it takes to look at things hard enough to recognize
similarities and connections and relationships.

Oh, sure, I know that everybody espouses the
virtues of kicking back and letting the world go by.

But people who let the world go by simply let the
world go by.

They don't make a mark. They don't make a
difference. They don't come up with ideas.

And that's what we're trying to do, isn't it? Come
up with ideas?

OK then, listen to me—when you forget about one
thing, start working on another thing.

In advertising, writers and art directors do this
whenever they can. When they're having trouble
coming up with, say, ideas for a television commercial
for a motorcycle and it isn't due until next week, they
put that assignment aside and start working on ideas
for a newspaper ad on cheese, or for a billboard on
a bank. A couple of days later they come back to the
motorcycle assignment and, magically, ideas fill the
room.

But what if you don't have another project to
work on?

Then get one.

The secret is to switch gears; to let your
unconscious work on the problem that's giving
you trouble, while your conscious mind works on
something else; to "sleep on" one problem while you
start working on another.

Carl Sagan did that. When he got stuck on one project, he moved on to the next, allowing his unconscious to go to work. "When you come back," he wrote, "you find to your amazement, nine times out of ten, that you have solved your problem—or your unconscious has—without you even knowing it."

So did Isaac Asimov. "When I feel difficulty coming on," he wrote, "I switch to another book I'm writing. When I get back to the problem, my unconscious has solved it."

But again: Keep working on something. Get another project and work on it.

Don't think that you've got to give your brain a rest. You don't. It's not a muscle that gets fatigued.

Besides, your unconscious doesn't know or care whether it's working on a project that might change the world or on solving the latest trashy whodunnit. It works just as hard regardless.

That's one of the reasons busy people get a lot done and can always handle another project—they've learned to focus their efforts on meaningful projects.

And they've learned to let much of their work "proceed underground."

For here is a super truth:

The more you do, the more you do;
the less you do, the less you do.

You know it's true. You know that one weekend you make a list of things you want to get done around the house and all of a sudden you get busy and you discover that you've got a lot of things to do and you get them all done. Another weekend you sit on your ass and watch the world go by and don't do a damn thing.

Work creates work. Effort creates effort.

And ideas create ideas.

After all, you have to think about something so why not think about some other idea or problem or project?

And if after a while the solution to the original problem doesn't come to you while you're shaving or milking a cow or getting on a bus, start back to work on it. When you do, you'll probably see roads that were not there before; closed doors will be open, barriers will be down; you'll have new insights and feel new hope and see new relationships and connections and structures and possibilities.

And that's when the idea will hit.

Wham.

"Ah," you'll say. "Why didn't I think of that before?"

15.

Put the Idea into Action

Even if you're on the right track,
you'll get run over if you just sit there.

Will Rogers

A vice president
in an advertising
agency is a "molehill man."
A molehill man is a pseudobusy executive who comes to
work at 9 a.m. and finds a molehill on his desk. He has
until 5 p.m. to make this molehill into a mountain. An
accomplished molehill man will have his mountain finished
even before lunch.

Fred Allen

When I was kidnapped, my parents snapped into action.
They rented out my room.

Woody Allen

As we discussed in chapter 7, you must screw up your courage and tell somebody about your idea.

And if it meets with yawns or jeers, you must press on.

But what happens when it's met with applause?

George Ade was a prolific writer in the early part of the 1900s. I once read an interview of his mother by a man who was not an admirer of her son's work, and he was indelicate enough to ask her about George's alleged capricious style and wobbly structure and shallow characterizations.

Finally Mrs. Ade had enough. "Oh, I know that many people can write better than George does," she said. "But George does."

"George does."

It's one of the finest things anybody ever said.

In two words it crystallizes what happens with so many people (me included), namely: They get an idea, they tell some people about it, the people all say, "Wow, that's great!" and then they go on to something else and never do anything more about the idea they told people about.

I think the reason is that "Wow, that's great!" is reward enough. It gives you the nice warm glow that comes from knowing you got a really good idea, that everybody thinks you're a whiz.

But if nothing else happens with your idea, if it doesn't help someone, if it doesn't save or fix or create something, if it doesn't make something better or solve some problem, what good is it really?

The truth is: There is no difference between (a) having an idea and not doing anything with it and (b) not having an idea at all.

So if you don't plan to do anything with your idea once you have it, don't come up with the idea in the first place. It's just a waste of time and energy.

And when you do have an idea, either (a) don't tell anybody about your idea or (b) don't let "Wow, that's great!" be enough.

OK? We've agreed? If you have an idea, you promise to screw up your courage once more and take the next step? Good.

OK, here are some things that might help:

Start Right Now

Will your enthusiasm for your idea be greater or lesser tomorrow? Then why wait?

Ralph Waldo Emerson said: "Nothing great is ever accomplished without enthusiasm." And the more enthusiasm the better.

Besides, most of the time, waiting to start anything is wrong.

Get it going now. Once you break the inertia and start it rolling, an idea takes on a life of its own and starts going into areas you never dreamed it would fit in; it creates opportunities; it bowls over barriers and leapfrogs objections and overwhelms logic.

If You're Going to Do It, Do It

If you don't commit yourself to making your idea work, you'll probably be looking back weeks or months from now saying, "Gee, if only I had done" this or that.

One of the best ways to commit yourself is to commit your money. Take some money out of your savings account or borrow some money from your brother-in-law, open up a checking account with it under the name of your idea, and spend some of it on something you need to do to get your project going.

That's commitment. And commitment creates action.

Give Yourself a Deadline, the Shorter the Better

It's amazing what you can get done if you know you must get it done.

Thomas Edison often predicted he'd invent something or other by such and such a time.

F. R. Upton, one of his closest associates, said: "I have often thought that Edison got himself into trouble purposely, by premature publication . . . so that he would have a full incentive to get himself out of trouble."

I used to do this all the time in developing advertising. "We'll get three more ideas," I'd say to my partner at noon, "and then we'll break for lunch."

Sure enough we'd get three more ideas. Lunch is essential.

Make a List of the Things You Have to Do If You Are to Put Your Idea into Action

Then every day do at least one thing on that list.

If you feel you're in a bit over your head because your idea is outside your area of expertise, go to the library or surf the Internet and read up on that area. Or ask someone about it. Or take a college course on it.

If you need a drawing made, get it made.

If you need a patent attorney, call one. The idea for barbed wire had been around for years, but Joseph Glidden did something about it. In 1873 he applied for a patent for a two-strand design called Winner and made millions.

If you have to write a brochure, start writing.

If you have to learn to play the guitar, put down this book and phone a guitar teacher.

If you have to . . . ah, you get the idea.

But remember: Do something about your idea every day. Open your computer or your folder or your notebook and do something. Every day. Even if it's only to review what you did yesterday, do it.

At the end of a month you'll be surprised at how much you've accomplished. At the end of a year you'll be astounded.

"Burn Your Boats"

Julius Caesar and other generals used this technique when they invaded a foreign country. It was a dramatic demonstration to their troops that since retreat was impossible, they must either conquer the country or die; there were no alternatives, there were no excuses.

What excuses will you use if you fail? Burn them.

You didn't have enough money? OK, borrow some money. Now you no longer have the lack-of-money excuse to fall back on.

You didn't have enough time? OK, burn that boat: Get up an hour or two early every morning and work on your idea.

You didn't know enough? OK, learn.

"Burn your boats."

If You Have Trouble Selling Your Idea to Somebody Else, Do It Yourself

Thomas Adams tried to sell to a major company his idea of a gum that people would chew. They turned him down. So he made and sold it himself and started a whole new industry. His four sons each inherited a fortune.

Walt Whitman couldn't find anybody to publish his *Leaves of Grass*, so he published it himself. e. e. cummings did the same thing with *No Thanks*. So did Mark Twain with *Huckleberry Finn*; and John Grishham with *A Time to Kill*; and Irma Rombauer with *The Joy of Cooking*; and Richard Bolles with *What Color Is Your Parachute?* And so did thousands of others with their books.

After Marion Donovan invented the disposable diaper, she tried for years to sell her invention to established manufacturers. No one wanted it. So she started her own company.

Steve Jobs and Steve Wozniak went to Atari and said: "'Hey, we've got this amazing thing, even built it with some of your parts, and what do you think about funding us? Or we'll give it to you. We just want to do it. Pay our salary, we'll come to work for you.' And they said, 'No.' So then we went to Hewlett-Packard, and they said, 'Hey, we don't need you. You haven't got through college yet.'" So the two Steves decided to make Apple computers themselves.

Do you really believe in your idea?

Then why let people who haven't thought about it and worked on it a tenth as much as you have put the kibosh on it?

Attack.

Stay with It

Everybody has a story about getting an idea ("I've got this idea, see?") for an investment or an invention or a new product or a new service or some new use for an existing product or some new way to use an existing service or a way to save money or an event or a promotion or a discovery or a screenplay or a book or a parlor game or a video game or a home video or a computer program or a real estate opportunity or a get-rich-quick scheme like selling used coffee grounds combined with pulverized orange rinds as an aromatic fertilizer for indoor plants but, alas, they never do anything with the idea and somebody else gets all the credit and makes a fortune off it.

I certainly have such a story. A number of them, in fact. You probably do too.

Here are a couple of famous ones:

James Clerk Maxwell predicted and mathematically formulated the transmission of radio waves. But he was a mathematician, and like a true mathematician, once he figured the thing out, he considered himself finished.

Robert Hooke probably discovered the law of gravity before Newton formulated his law of gravitation, and theories of light and color before Newton's book on optics. But he never followed through with either discovery.

The first real sewing machine (one with an

overhanging arm, a perpendicular action, an eye-pointed needle, and a feed similar to most machines today) was invented and patented in 1790 by Thomas Saint. Unfortunately for Mr. Saint, he never built even one of his machines. Forty years later, Barthelemy Thimonnier independently invented and built a similar machine, and the era of modern sewing machines began.

Here's a poster my first boss, Bud Boyd, kept on his wall:

Nothing in the World Takes the Place of Persistence

Talent doesn't—
nothing is more common than
unsuccessful men with talent.

Wealth doesn't—
the born-rich who die poor are legion.

Genius doesn't—
unrewarded genius is almost a proverb.

Education doesn't—
the world is full of educated derelicts.

Luck doesn't—
her fickleness has toppled kings.

Persistence and Determination Alone Are Omnipotent

"More often than not," Bud said, "people don't fail; they stop trying."

Don't stop. Stay with it.

Make a copy of Bud's poster and put it on your wall.

Ten, fifteen, twenty years from now, chances are the things you'll regret most won't be the dumb things you did during that time. They'll be the things you didn't do—the chance you didn't take, the opportunity you didn't seize, the idea you didn't stay with.

Take it.

Seize it.

Stay with it.

Give Yourself Reasons

It took me three years to write this book. It took me so long because I followed none of the rules I just gave you except, marginally, the one about staying with it.

Truth to tell, many months I wrote nothing.

Furthermore, when I did write, I wrote more slowly than a tree grows.

And although it may not look it, for every three sentences I wrote, I threw away two; and every sentence I kept, I rewrote three times and repunctuated four.

But I also gave myself reasons to stay with it, to write and finish it.

The reasons were many: money, recognition, pride, stubbornness, curiosity, fun, the desire to help, the thrill of the reaching sail.

Mostly though, I stayed with it because I knew it would give me a chance to work again with someone I missed working with—the illustrator of this book.

Find reasons that will motivate you to put your idea into action.

One could be as simple as the satisfaction you feel when you finish what you start. Or the pride on your spouse's face. Or the buzz your finished project will create. Or one of the reasons that propelled me.

Make a list of your reasons. Put the list on the wall next to Bud's poster where you can see both of them every day.

Then don't simply look at them every day.

See them every day.

Notes

Introduction: What Is an Idea? 1

The Charles Schultz quote is from a *Peanuts* comic strip, January 1982.

Mark Twain, *The Portable Mark Twain* (New York: Viking Press, 1963).

The Lily Tomlin quote is from http://quotations.home.worldnet.att. net.

Page 2

A. E. Housman as quoted by Arthur Koestler, *The Act of Creation* (New York: Macmillan, 1964).

The dictionary definitions are from *American Heritage Dictionary*, *Webster's Third New International Dictionary*, and *Webster's Seventh New Collegiate Dictionary*.

Page 3

Marvin Minsky, *The Society of Mind* (New York: Simon and Schuster, 1988).

Page 4

James Webb Young, *A Technique for Producing Ideas* (Chicago: Advertising Publications, 1951).

Page 5

Jacob Bronowski, *A Sense of the Future* (London, England: MIT Press, 1977).

Jacques Hadamard as quoted by Arthur Koestler, *The Act of Creation*.

Page 6

T. S. Eliot as quoted by Richard Lederer, *The Miracle of Language* (New York: Pocket Books, a division of Simon and Schuster, 1991).

Bronowski, *A Sense of the Future.*

Robert Frost as quoted in *Contemporary Quotations*, compiled by James B. Simpson (Boston: Houghton Mifflin, 1988).

Francis H. Cartier as quoted by Laurence Peter, *Peter's Quotations*. (New York: Bantam, 1979).

Page 7

Nicholas Negroponte, *Being Digital* (New York: Vintage, 1996).

Koestler, *The Act of Creation.*

Page 8

The creative steps of Helmholtz are from James F. Fixx, *More Games for the Super-Intelligent* (New York: Popular Library, 1972).

The creative stages of Moshe F. Rubinstein are from James F. Fixx, *Solve It* (New York: Fawcett Popular Library, 1978).

1. Have Fun 15

Mary Pettibone Poole, *A Glass Eye at a Keyhole* (Philadelphia: Dorrance, 1938).

Guy Davenport as quoted by William Buckley in the *New York Times Book Review*, 24 March 1977.

Oscar Wilde as quoted by Robert Byrne, *The 637 Best Things Anybody Ever Said* (New York: Fawcett Crest, 1982).

Page 17

David Ogilvy, "How to Run an Advertising Agency," advertisement.

Roger von Oech, *A Kick in the Seat of the Pants* (New York: Harper & Row, 1986).

Paul Valéry as quoted in *The New International Dictionary of Quotations* (New York: E. P. Dutton, 1986).

The Issac Asimov quote is from http://en.wikipedia.org.

Notes

Page 18

Woody Allen, "Selections from the Allen Notebook" in *Without Feathers* (New York: Random House, 1975).

Damon Runyon as quoted by Byrne, *The 637 Best Things Anybody Ever Said*.

Ring Lardner, *The Young Immigrants* (New York: Scribner, 1962).

Page 19

The Hutchins and Lipman references are from David Wallechinsky and Irving Wallace, *The People's Almanac* (New York: Doubleday, 1975).

Page 22

Jerry Greenfield as quoted in *Rolling Stone*, 9 July 1992.

Tom J. Peters as quoted in the *Los Angeles Times*.

2. Be More Like a Child   25

Ralph Waldo Emerson, *Nature* (Boston: Beacon Press, 1991).

Fred Allen and George Bernard Shaw as quoted by Byrne, *The 637 Best Things Anybody Ever Said*.

The Sam Levenson quote is from http://www.brainyquote.com.

Page 26

Charles Baudelaire as quoted on http://www.brainyquote.com/quotes.

Gary Zukav, *The Dancing Wu Li Masters* (New York: William Morrow, 1979).

Page 27

Jean Piaget as quoted by Eugene Raudsepp, *Creative Growth Games* (New York: Jove Publications, 1977).

Page 28

J. Robert Oppenheimer as quoted by Marshall McLuhan, *The Medium Is the Massage* (New York: Random House, 1967).

Thomas Edison as quoted by Mike Vance and Diane Deacon, *Think Out of the Box* (Franklin Lakes, N.J.: Career Press, 1995).

The Will Durant quote is from a letter in a 1983 Jack Smith column in the *Los Angeles Times*.

Albert Einstein as quoted by John D. Barrow, *Theories of Everything* (Oxford, England: Clarendon Press, 1991).

Dylan Thomas, *The Poems of Dylan Thomas* (New York: New Directions, 1971).

Page 29

Robert M. Pirsig, *Zen and the Art of Motorcycle Maintenance* (New York: William Morrow, 1974).

Page 30

Carl Sagan as quoted in the *Los Angeles Times*, 4 December 1994.

Neil Postman as quoted by Raudsepp, *Creative Growth Games*.

3. Become Idea-Prone   35

Samuel Johnson as quoted by James Boswell, *Life of Johnson* (New York: Viking Penguin, 1979).

G. C. Lichtenberg and Kent Ruth as quoted by Peter, *Peter's Quotations*.

Page 38

The shaving mugs reference is from David Louis, *2201 Fascinating Facts* (New York: Random House Value, 1988).

The barbed wire reference is from *The Antique Trader*.

Page 39

Lincoln Steffens, *The Autobiography of Lincoln Steffens* (New York: Harcourt Brace Jovanovich, 1931).

Page 41

Émile Coué as quoted by Maxwell Maltz, *Psycho-Cybernetics* (New Jersey: Prentice-Hall, 1960).

Page 42

Norbert Wiener, *The Human Use of Human Beings* (Boston: Houghton Mifflin, 1950).

Notes

Joseph Heller as quoted by George Plimpton, *The Writer's Chapbook* (New York: Viking Penguin, 1989).

Thomas Edison as quoted by Maltz, *Psycho-Cybernetics*.

Page 45

Virgil as quoted by Anthony Robbins, *Unlimited Power* (New York: Simon and Schuster, 1986).

Page 46

Henry Ford as quoted by Roger von Oech, *A Kick in the Seat of the Pants*.

William James as quoted by Alfred Armand Montapert, *Distilled Wisdom* (New Jersey: Prentice-Hall, 1964).

Jean-Paul Sartre, *Existentialism and Human Emotions* (New York: Citadel Press, 1971).

Anton Chekhov as quoted by W. H. Auden, *A Certain World* (New York: Viking Press, 1970).

4. Visualize Success 51

Robert Frost as quoted by Barbara Rowes, *The Book of Quotes* (New York: E. P. Dutton, 1979).

Lily Tomlin as quoted by Robert Byrne, *The Other 637 Best Things Anybody Ever Said* (New York: Fawcett Crest, 1984).

Phyllis Diller as quoted by Byrne, *The 637 Best Things Anybody Ever Said*.

Page 54

The basketball and dart stories are from Maltz, *Pyscho-Cybernetics*.

5. Rejoice in Failure 59

The Gore Vidal quote is from *The Columbia World of Quotations* (New York: Columbia University Press, 1996).

H. L. Mencken, *A Mencken Chrestomathy* (New York: Knopf, 1949).

The Paul Newman quote is from Maureen Dowd, "Testing Himself" in the *New York Times*, 28 September 1986.

Leo Burnett, *A Tribute to Leo Burnett* (Chicago: Leo Burnett, 1971).

Page 62

Thomas Edison as quoted on: http://www.wisdomquotes.com

Jerry Della Femina, as quoted in *Modern Maturity*, March/April, 2002.

Page 64

The Chester F. Carlson, Bette Nesmith Graham, and Alfred Mosher Butts references are from: http://inventors.about.com/od/lstartinventions.

The James Russell reference is from http://www.popularmechanics.com/specials/features.

Page 65

The *Catch 22*, Dr. Seuss, *Sister Carrie*, *Chicken Soup for the Soul*, and *Zen and the Art of Motorcycle Maintenance* references are all from Dan Poynter's website: http://www.parapublishing.com.

6. Get More Inputs   67

Fletcher Knebel as quoted in *Contemporary Quotations* compiled by Simpson.

Ethel Watts Mumford as quoted by Peter, *Peter's Quotations*.

H. L. Mencken as quoted by Byrne, *The 637 Best Things Anybody Ever Said*.

Page 72

Jerry Della Femina, as quoted in *Modern Maturity*, March/April, 2002.

Page 74

Louis L'Amour as quoted by Vance and Deacon, *Think Out of the Box*.

7. Screw Up Your Courage   83

Franklin P. Jones and Woody Allen as quoted by Byrne, *The 637 Best Things Anybody Ever Said.*

Jamaican proverb as quoted by Peter, *Peter's Quotations.*

Page 84

Charles Brower as quoted in *Contemporary Quotations*, compiled by Simpson.

Page 85

Robert Grudin, *The Grace of Great Things* (New York: Ticknor & Fields, 1990).

Pages 86

The Richard Drew reference is from Vance and Deacon, *Think Out of the Box.*

The Joseph Priestley and Blaise Pascal references are from Louis, *2201 Fascinating Facts.*

The Alexander Graham Bell reference is from R. Keith Sawyer, author of *Explaining Creativity: The Science of Human Innovation*, as quoted in *Time* magazine, 16 January 2006.

The George Crum, Slinky Toy, and microwave oven references are from http://www.enchantedlearning.com/inventors.

Page 87

The pacemaker reference is from http://www.popularmechanics.com/specials.

The other "bad idea" references are from Wallechinsky and Wallace, *The People's Almanac.*

The Teflon, Krazy Glue, and Scotchguard references are from *Time* magazine, 13 February 2006, *People's Almanac*, or are common knowledge.

Page 88

Ray Bradbury, *Zen in the Art of Writing* (Santa Barbara: Capra Press, 1990).

The Wright Brothers reference is from R. Keith Sawyer, author of *Explaining Creativity: The Science of Human Innovation*, as quoted in *Time* magazine, 16 January 2006.

Page 90

Grudin, *The Grace of Great Things*.

8. Team Up with Energy   93

The Samuel Goldwyn quote is as quoted by Barbara Rowe, *The Book of Quotes* (New York: E. P. Dutton, 1979).

The Lily Tomlin quote is from http://quotations.home.worldnet.att. net.

Page 99

David Ogilvy, *How to Run an Advertising Agency*, advertisement, circa 1980.

9. Rethink Your Thinking   101

Bertrand Russell as quoted by A. Flew, *Thinking about Social Thinking* (Buffalo, N.Y.: Prometheus Books, 1995).

James Thurber and Martin H. Fischer as quoted by Peter, *Peter's Quotations*.

Page 102

The Albert Einstein reference is from Eugene Raudsepp, *How Creative Are You?* (New York: Perigee Books, 1981).

Pages 103

The David Hilbert reference is from George Gamow, *One, Two, Three . . . Infinity* (New York: Viking Press, 1947).

The Kelvin, Freud, and Newton references are from Koestler, *The Act of Creation*.

Page 106

Edward de Bono, *Lateral Thinking* (New York: Harper & Row, 1970).

Page 114

Rollo May, *The Courage to Create* (New York: W. W. Norton, 1975).

Joseph Heller as quoted by George Plimpton, *The Writer's Chapbook*.

Leonardo da Vinci as quoted by W. H. Auden and Louis Kronenberger, *The Viking Book of Aphorisms* (New York: Viking Press, 1992).

Page 115

Duke Ellington as quoted by May, *The Courage to Create*.

The Walter Hunt reference is from Wallechinsky and Wallace, *The People's Almanac*.

John Dryden as quoted by Bronowski, *A Sense of the Future*.

May, *The Courage to Create*.

10. Learn How to Combine   117

Bum Phillips as quoted by Byrne, *The 637 Best Things Anybody Ever Said*.

Steve Allen as quoted in *Contemporary Quotations*, compiled by Simpson.

Rousseau et al. as quoted by J. M. and M. J. Cohen, *The Penguin Dictionary of Modern Quotations* (New York: Penguin Books, 1980).

Page 124

Roger von Oech, *A Whack on the Side of the Head* (New York: Warner Books, 1983).

Page 125

The James J. Ritty reference is from David Wallechinsky and Irving Wallace, *The People's Almanac #2* (New York: A Bantam Book, 1978).

Charles Darwin as quoted by Robert B. Downes, *Books That Changed the World* (New York: Mentor Books, 1956).

The Benjamin Huntsman reference is from James Burke, *Connections* (Boston: Little Brown, 1978).

The George Westinghouse reference is from the editors of American Heritage, *The Confident Years* (New York: American Heritage Publishing, 1969).

Page 126

The Jim Crocker reference is as quoted in *Time* magazine, 16 January 2006 by R. Keith Sawyer, author of *Explaining Creativity: The Science of Human Innovation*.

The Descartes, Oersted et al., and Kepler references are from Koestler, *The Act of Creation*.

Page 127

Thelonious Monk as quoted by Bill Bernbach in a speech given on 17 May 1980.

11. Define the Problem   131

Charles Schultz quote is from http://www.allgreatquotes.com/charles_schulz.

James Thurber and Gore Vidal as quoted by Byrne, *The 637 Best Things Anybody Ever Said*.

Picasso as quoted by William Fifield, *In Search of Genius* (New York: William Morrow, 1982).

Page 132

Albert Einstein as quoted by Anne C. Roark in the *Los Angeles Times*, 29 September 1989.

Page 133

The Edward Jenner reference is from Edward de Bono, *New Think* (New York: Avon Books, 1967).

Koestler, *The Act of Creation*.

Jonas Salk as quoted by Bill Moyers, *A World of Ideas with Bill Moyers, PBS 1990* (New York: Doubleday, 1990).

12. Gather the Information   145

Will Cuppy, *The Decline and Fall of Practically Everybody* (New York: Holt, 1950).

Steve Kravitz and Thomas Edison as quoted by Byrne, *The 637 Best Things Anybody Ever Said*.

Notes

Page 150

Bud Robbins, "Looking for the Capo d'astro bar," advertisement.

Page 152

Bill Bernbach, in a speech given on 17 May 1980.

Page 154

Thomas Mann as quoted in *The Harper Book of Quotations* (New York: Harper Perennial, 1993).

13. Search for the Idea   157

Florynce Kennedy as quoted by Gloria Steinem in "The Verbal Karate of Florynce Kennedy," *Ms.*, March 1973.

Gene Fowler as quoted by Jon Winokur, *Writers on Writing* (Philadelphia: Running Press, 1986).

Page 158

Hal Riney, *ADWEEK*, February 1983.

Linus Pauling as quoted by von Oech, *A Kick in the Seat of the Pants*.

Page 163

The Andrew Wiles reference is from *People Weekly*, 27 December 1993–3 January 1994.

The Gatling reference is from Wallechinsky and Wallace, *The People's Almanac*.

The Nikola Tesla reference is from Wallechinsky and Wallace, *The People's Almanac #2*.

14. Forget about It   165

Publilius Syrus as quoted by Byrne, *The 637 Best Things Anybody Ever Said*.

Eric Morecambe and Ernie Wise as quoted by Fred Metcalf, *The Penguin Dictionary of Modern Humorous Quotations* (London: Penguin Books, 1987).

Italo Svevo as quoted by Peter, *Peter's Quotations*.

Page 166

Andy Rooney, *ADWEEK*, 7 February 1983.

Page 167

Hermann von Helmholtz as quoted by Fixx, *More Games for the Super-Intelligent*.

The Albert Einstein reference is from *Creativity*, edited by Paul Smith (New York: Hastings House, 1959).

Grant Wood as quoted in *Contemporary Quotations*, compiled by Simpson.

The Henri Poincaré reference is from *Creativity*, edited by Smith.

Bertrand Russell, *The Conquest of Happiness* (New Jersey: Garden City Publishing, 1933).

The C. G. Suits reference is from Maltz, *Pyscho-Cybernetics*.

Page 168

May, *The Courage to Create*.

Lloyd Morgan as quoted by Koestler, *The Act of Creation*.

Philip Goldberg, *The Babinski Reflex* (Los Angeles: Jeremy P. Tarcher, 1990).

Page 170

Carl Sagan as quoted in the *Los Angeles Times*, 4 December 1994.

Isaac Asimov as quoted by Winokur, *Writers on Writing*.

15. Put the Idea into Action 173

Will Rogers as quoted by Vern McLellan, *Wise and Wacky Wit* (Wheaton, Ill.: Tyndale House Publishers, 1992).

Fred Allen as quoted in *Contemporary Quotations*, compiled by Simpson.

Woody Allen as quoted by Byrne, *The 637 Best Things Anybody Ever Said*.

Page 175

Ralph Waldo Emerson as quoted by John Bartlett in *Familiar Quotations* (Boston: Little Brown, 1955).

Page 177

F. R. Upton as quoted by Matthew Josephson, *Edison* (New York: McGraw-Hill Paperbacks, 1959).

The Joseph Glidden reference is from Owen Edwards, *Elegant Solutions* (New York: Crown Publishers, 1989).

Page 178

The Thomas Adams reference is from Wallechinsky and Wallace, *The People's Almanac #2*.

Page 179

The self-publishing references are from: http://www.llumina.com.

The Marion Donovan reference is from http://www.enchantedlearning.com/inventors.

The Apple computer reference is from http://www.goofups.com/quotes.

Page 180

The James Clerk Maxwell and Robert Hooke references are from pamphlets issued by the Karpeles Manuscript Library.

The sewing machine reference is from http://www.usgennet.org/usa/topic/preservation/science/inventions.

Index

A Technique for Producing Ideas
(Young), 8
A Time to Kill (Grishham), 179
A Whack on the Side of the Head
(von Oech), 124
accomplishments, deadlines, 115
achievement, creative, 90
action, putting ideas into, 174–175
Adams, Thomas, 178–179
Ade, George, criticism of his writing
style, 174
advertising
agencies, 17
nature of, 87
Aeolian Piano Company, 150–151
age, mental image, 32
airplanes, the first, 88
Allen, Fred
boring adults, 25
the molehill man, 173
Allen, Steve, allergies and asthma,
117
Allen, Woody
believing in God, 18
fear of dying, 83
taking action, 173
America, discovery of, 87
analogues, using comparisons,
118–119
answers
easy, 41
alternative, 162
Apple Computer, 179
architecture, new designs, 120

art
children's, 21
individual pursuits, 94
new techniques, 119–120
Asimov, Isaac
exciting phrases, 17
letting your unconscious solve a
problem, 170
Ask, researching on the Internet,
153
assignments. *See* exercises
assumptions
changing, 138–141
limiting choices, 109–110
astronauts, NASA recruit selection,
63
astronomy, modern, 126
athletics, new techniques, 120
atom, image of an, 103
attention, attracting, 36–37
attitude, performance, 46
attorneys, patent, 177

bacon, advertisement for, 146–150
bar, Capo d'astro, 151
barbed wire, Winner, 177
Bartley, Bill
Knudsen yogurt billboard, 96
leadership images, 104
baseball, center fielding, 53–54
basketball, free throw exercise,
54–55
Baudelaire, Charles, childhood
genius, 26

Bean, Bob, observation game, 76–78
Becquerel, Henri, 87
Beethoven, Ludwig van, 120
Bell, Alexander Graham, 86
Ben & Jerry's Ice Cream, 22
Bernbach, Bill, need of a
    springboard, 152
Berrett-Koehler, Steven Piersanti,
    65
billboards, Knudsen yogurt, 96
bisociation, 7, 17–18
Black Plague, burial of victims, 138
boats, burning your, 178
Bohr, Niels, 103
Bolles, Richard, What Color Is Your
    Parachute?, 179
boss, teamwork presentations to, 99
boundaries
    crossing disciplinary, 124
    unconscious, 109–110
Boyd, Bud
    poster on persistence, 181
    bacon advertisement research,
        146–150
Boyd, Johnny-Boy, accident prone,
    36
Bradbury, Ray, writing short stories,
    71, 88
brainstorming, teamwork, 99
brakes, air, 125
breaks
    fun, 21
    taking, 168
Bronowski, Jacob, creative activity,
    5, 6
Brower, Charles, fragility of ideas,
    84
Brown, Charlie, 153
Brown, Norm, knowing your
    destination, 132
Burnett, Leo, making mistakes, 59
Butts, Alfred Mosher, 64

Caesar, Julius, no retreat, 178
Camus, Albert, 85
Capo d'astro bar, 151
Carême, Antonin, 121
Carlson, Chester F., 64
cars, design of, 31
Cartier, Francis H., discovering
    relationships, 6
cash register, 125
Catch 22, 65
chair
    drawing a, 78–80
    new designs, 120
Chamberlain, Wilt, free throws, 43
chances, taking, 62–63, 126–127
change
    enabling, 46–47
    fear of, 85–86
    personal mental image, 48
Chekov, Anton, belief in self, 46
chewing gum, 178–179
Chicken Soup for the Soul, 65
clock, alarm, 19
cogito, 5
Columbus, Christopher, 87
comfort, performance level, 44–45
commercials
    heated shaving cream, 97
    Knudsen yogurt, 96
    Smokey Bear, 32, 39–40
commitment, to making it work, 176
comparisons
    analogues, 118–119
    evaluating solutions, 61
    of similar products, 150–151
concepts
    capturing, 3–4
    infinity, 103
    insight, 4
    mind over body, 47
conditioning, problem solving 10
confidence
    childhood, 27

following rejection, 65
performance, 43–45
connections
  new, 89–90
  quick, 72
consciousness, focusing attention
  on a subject, 153–154
constraints, assuming, 109
conversations, between fruit and
  vegetables, 32
cookbooks
  breaking the rules, 121
  proliferation of, 38
cooking
  creating recipes, 115
  creation of, 18
  learning from failure, 60
cookoff, office chili, 22
Coover, Harry, 87
Corby, Larry
  about the illustrator, 201–202
  capturing ideas, 42
Coué, Émile, easy answers, 41
courage
  finding, 84–85
  taking action, 174–175
Craig, Jean, targeting the right
  market, 97–98
creativity
  basis of, 17–18
  and failure, 62
  humor, 19
  mental, 9
  unlikely juxtapositions, 7
  within a limited framework, 114
criticism, taking honest, 83
Crocker, Jim, 126
Crum, George, 86
cummings, e.e., 120
  No Thanks, 179
Cuppy, Will, Aristotle's teachings on
  brain function, 145
Curie, Madame, 86

curiosity, insatiable, 30–31, 68–69
current
  alternating, 163
  electric, 86

da Vinci, Leonardo, disciplined
  creativity, 114
Dali, Salvador, 18
darts
  mentally throwing, 55
  playing, 21
Darwin, Charles, 19, 125
Daugerre, Louis-Jacques-Mande, 87
Davenport, Guy, Goethe's humor, 15
days, dress-up, 22
de Bono, Edward, lateral thinking,
  106–109
deadlines
  accomplishments, 115
  setting, 176–177
Della Femina, Jerry
  failure and creativity, 62
  quick connections, 72
Descarte, René, 126
design, automobile, 31
desserts, breaking the rules, 121
diapers, disposable, 179
Diller, Phyllis, beauty parlors, 51
discoveries
  implementing, 180–181
  from bad ideas, 86–87
  from breaking the rules, 119–121
  from changing the question, 133
  from unrelenting effort, 163
  See also inventions
Donovan, Marion, 179
donuts, increasing sales, 139–140
dots, connecting nine, 110
doubt, self-image, 44–45
downsizing, free time, 49
Dr. Seuss, 65
Drew, Richard, 86
drivers, race car, 60

Dryden, John, rhymed verses, 115
Durant, Will, cosmic truth, 28

Eames, Charles, 120
Edison, Thomas
  deadlines, 176–177
  extent of human knowledge, 145
  on failure, 62
  ideas in the air, 42
  inventing the lightbulb, 88
  the mind of a child, 28
Einstein, Albert, 28
  getting good ideas, 167
  images, 102–103
  problem statements, 132
Einstein, Cliff
  generating ideas, 158
  SOAPRIZE, 95
electromagnetism, field of, 126
elements, combining old, 4, 69, 71
elevators, waiting for, 134–136
Eliot, T. S.
  creating a new whole, 6
  creativity within a framework,
    114–115
Ellington, Duke, music for specific
  instruments, 115
Emerson, Ralph Waldo
  enthusiasm, 175
  nature of a child, 25
employees, solving management
  problems with lateral thinking,
  106–109
enthusiasm, accomplishments, 175
Essay on Population (Malthus), 125
Eureka Effect, 168
excuses, eliminating, 178
exercises
  basketball free throw, 54–55
  classroom, 111–113
  half of thirteen, 160–162
  paper airplane, 111
  Ping Pong ball, 111–112

Swiss Army knife, 40–41
  the 2x2x2 wood block, 159
  See also problems
experiences, opening yourself to
  new, 72–74
explanations, using analogues,
  118–119

failure
  fear of, 63–64
  refusing to accept, 64
  searching far enough, 60
  self-image, 45
  success, 61
fairs, arts and crafts, 21
family day, at work, 21
Faraday, Michael, 126
Farmer, Fanny, 121
fear
  criticism, 84–86
  failure, 63–64
fences, swinging for, 89–90
Fischer, Martin H., reaching a
  conclusion, 101
Fleming, Alexander, 87
Florida, spring break behavior,
  136–137
focus, subconscious attention, 153
Ford, Henry,
  changing the question, 133
  positive attitude, 46
  worker's wages, 120
Fosbury, Dick, 120
Foster, Jack, about the author,
  199–200
Fowler, Gene, writing, 157
framework, solving problems,
  113–114
freedom, taking chances, 62
Freud, Sigmund, 103, 120
friends, teamwork, 94
Frost, Robert
  mental abilities, 51

Index

remembering associations, 6
frustration, using distraction to
    reduce, 134–136
fun
    idea conditioning, 16
    workplace ambiance of, 19–22

Galvani, Luigi, 86
galvanometer, mirror, 103
games
    observation, 76–78
    Scrabble, 64
    "What If?", 121–123
    white horse, 75
gasoline, antiknock, 86
Gatling, Richard, 163
Gaudí, Antoni, 120
gears, switching, 169
genius, childhood, 26
geometry
    analytical, 126
    Euclidean, 120
Gide, André, reading about new
    subjects, 72
gifts, great, 97–98
Glidden, Joseph, 177
goals
    staying with your, 182–183
    visualization, 52–54
Goethe, humor of, 15
Gogolak, Pete, 120
Goldberg, Philip, *The Babinski Reflex*, 168
Goldwyn, Samuel, truth, 93
golf, tournament leaders, 43
Goodyear, Charles, 86
Googling, on the Internet, 153
Graham, Bette Nesmith, 64
gravity
    law of, 180
    theory, 18
Greatbatch, Wilson, 87
Greenfield, Jerry, fun at work, 22

Grishham, John, *A Time to Kill*, 179
groceries, delivering to the
    customer, 133
Grudin, Robert
    creative achievement, 90
    The Grace of Great Things, 85
guns, machine, 163
Gutenberg, Johannes, 18, 125

Hadamard, Jacques, invention, 5
handicapped, hiring the, 108–109
Harvey, William, images, 102
Heller, Joseph
    focusing imagination, 114–115
    ideas in the air, 42
Hemingway, Ernest, 85
Hilbert, David, concept of infinity,
    103
hockey, hallway, 21
Hoffer, Eric, researching subjects,
    154
Hooke, Robert, 180
Hopkins, Gerard Manley, 120
Housman, A.E., recognizing
    objects, 2
Hubble telescope, 126
Huckleberry Finn (Twain), 179
humor, creativity, 17–19
Hunt, Walter, 115
Huntsman, Benjamin, 125
Hutchins, Levi, 19

ideas
    being possessed by, 154
    capturing, 42
    defined, 2
    expectation of finding, 38
    frequency of, 36
    generating multiple, 89, 158–159
    producing, 11
    recognizing, 2–4
    visualization, 104

illustrations
  being childlike, 24
  defining the problem, 130
  digging up ideas, 128–129
  failure, 58
  gathering information, 144–145
  generating energy, 92–93
  getting into condition, 13
  having courage, 82
  having fun, 14–15
  incubating an idea, 164
  multiple inputs, 66
  raining ideas, 34–35
  rethinking, 100–101
  searching for ideas, 156–157
  selecting the best solutions, 172–
    173
  unusual combinations, 116
  visualizing success, 50
  wondering about a question, xiv
images, thought process, 102–105
immunology, 86
impasse, overcoming an, 168–169
inaction, eliminating excuses for,
  178
incubation, subconscious, 8–9
inertia, breaking, 176
infinity, concept of, 103
information
  data gathering, 146–150
  gathering background, 177
  knowing when you have enough,
    163
  resources, 152–154
insight, mental, 4, 9
inspiration, illumination, 8–9
instinct, sublimation of, 103
intelligo, 5
Internet, researching on, 73,
  152–153
invention, father of, 17
inventions
  great gift, 97–98

from a refusal to accept failure,
  64–65
from crossing disciplinary
  boundaries, 125–126
from image comparisons, 103
from specific needs, 115
selling it yourself, 178–179
See also discoveries

James, Richard, 87
James, William, attitude adjustment,
  46
Jenner, Edward, 133
Jobs, Steve, 179
Johnson, Samuel, lacking ideas, 35
Jones, Franklin P., taking criticism, 83

Kelvin, Lord, 103
Kennedy, Florynce, inaction, 157
Kepler, Johannes, 89, 126
Kettering, Charles, 86
Kirkegaard, Søren, 85
Knebel, Fletcher, statistics, 67
knowledge, gaining specific, 152
Koestler, Arthur
  asking the right questions, 133
  bisociation, 17–18
  solving problems, 42
  The Act of Creation, 7
Kravitz, Steve, rejecting alternative
  solutions, 145
Krazy Glue, 87

L'Amour, Louis, opening yourself to
  new experiences, 74
Lardner, Ring, humorous commands,
  18
Latin proverb, finding alternatives,
  157
Leaves of Grass (Whitman), 179
lessons, learning from failure, 61
Levenson, Sam, hereditary insanity,
  25

Index

Lichtenberg, G.C., real genius, 35
lightbulb, inventing the, 88
limitations, assumptions, 109, 113
limits, solution framework,
    113–114
Lipman, Hyman L., 19
Lippershey, Hans, 87
Liquid paper, 64
lists, to do action, 177–178
literature, analogues, 119
Lobachevsky, Nicolay Ivanovich,
    120
locks, comparative images, 104
logic, lateral thinking, 106–109
looking, vs. seeing, 75–76

Malthus, Thomas, *Essay on
    Population*, 125
Mann, Thomas, being possessed by
    an idea, 154
Mars, the orbit of, 89
Master locks, comparative images,
    104
matches, friction, 87
Maxwell, James Clark, 180
May, Rollo
    sources of inspiration, 168
    The Courage to Create, 114
    writing poetry, 115
Mays, Willie, center fielding, 53–54
medicine, new treatments, 120
meetings, in the park, 21
memory, observation, 80
Mencken, H.L.
    failure, 59
    human knowledge, 67
methods, idea production, 8–9
mind, altering your self image,
    47–48
mindset, conditioning to solve
    problems, 10
Minsky, Marvin, *The Society of
    Mind*, 2–3

money
    committing resources, 176
    increasing saving, 141
Monk, Thelonious, taking chances,
    127
moon, motion of the, 103
mop, inventing the, 19
Morecambe, Eric and Ernie Wise,
    on forgetting, 165
Morgan, Lloyd, waiting for answers,
    168
motivation
    finding, 162–163
    satisfaction, 183
Mumford, Ethel Watts, knowledge
    and power, 67
music, creating, 64, 120, 127

natural selection theory, 19, 125
negativity, avoiding, 99
Negroponte, Nicholas, creativity, 7
Newman, Paul, epitaph, 59
Newton, Sir Isaac, 18, 103
Nietzsche, Friedrich, 85
No Thanks (cummings), 179
notebook, keeping a personal, 81
now, knowing about, 30

observation
    conscious, 75
    the game of, 76–78
    memory, 80
Oersted, Hans Christian, 126
Ogilvy, David, 120
    making work fun, 17
    spreading gloom, 99
operations, household, 31
Oppenheimer, J. Robert, childhood
    sensory perception, 28
opportunity
    recognizing, 39
    rejection as an, 88
Osborne, Alex, brainstorming, 99

ovens, microwave, 87
pacemaker, 87
packaging, product, 31
Pareto, Vilfredo, *Speculators* and *Rentiers*, 37
park, meetings in the, 21
Pascal, Blaise, 86
Pasteur, Louis, 86, 120
patent attorneys, 177
patents
    new products, 38
    sewing machines, 180–181
    Winner barbed wire, 177
Pauling, Linus, getting good ideas, 158–159
Peanuts, Charlie Brown's tongue, 153
pencils, with erasers, 19
penicillin, 87
perception, childhood sensory, 28
performance, confidence, 43–45
persistence, omnipotence of, 181–182
perspective
    a cat's, 32
    childhood, 28–29
    playing "What If?" to change, 121–123
Peters, Tom J., fun at work, 22
Phillips, Bum, Don Shula's coaching prowess, 117
photography, 87
Piaget, Jean, being more creative, 27
pianos, researching, 150–151
Picasso, Pablo
    breaking rules, 120
    and computers, 131
pictures
    baby, 21
    thinking in images, 104
Ping-Pong ball, class exercise, 111–112

Pirsig, Robert, *Zen and the Art of Motorcycle Maintenance*, 29
Plunkett, Roy, 87
poetry, breaking the rules, 120
Poincaré, Henri, taking a break, 167
police, Florida spring break problems, 136–137
Poole, Mary Pettibone, laughter, 15
posters, omnipotence of persistence, 181–182
Postman, Neil, effect of school on children, 30
potato chips, 86
Predator of the Universe: The Human Mind (Wakefield), 9
presentations, to the boss, 99
press, printing, 18
Presume, Dr. Livingston I, 117
Price, Ralph, handling rejection, 61, 87–88
Priestley, Joseph, 86
printing press, 125
problems
    changing the definition, 134
    connecting nine dots, 110
    forgetting about, 166–171
    scientific method of solving, 8–9
    solving with "What If?", 121–123
    stating correctly, 132–133
    unconsciously solving, 170
    using analogues to solve, 118–119
    using lateral thinking to resolve, 106–109
    See also exercises, solutions
procedures, improving workplace, 31
production, improving, 141
products, researching, 150–154
proverbs, Jamaican on courage, 83
publishing, self, 179
questions
    asking, 30–31, 138–141, 146–150

changing the, 141–142
scientific, 30
queue, use of a, 31

radio waves, transmission of, 180
radioacitivity, 87
radium, 86
rationalizations, staying with your
    goal, 182–183
Ray, Man, images, 103
reading
    changing your self-image, 48
    new experiences, 72–74
    research, 153
regrets, actions not taken, 182
rejection
    confidence following, 65
    fear of, 84
    as an opportunity, 88
relationships
    fun and work, 20–22
    identifying existing, 8
    overlooked, 152
    between unrelated objects, 30
    between work and effort, 171
relativity, theory of, 28
relaxation, mental, 168–169
Rentiers, 37
research, surfing the Internet,
    152–153
Research Quarterly, improving
    performance, 54–55
resources, committing, 176
restrictions
    assuming, 109
    unconscious boundaries, 110
resumes
    evaluating, 62–63
    magazine ad, 69–71
retention, employee, 107–109
rewards, new ideas, 33
Richardson, Owen Willans, 126
Riney, Hal, the creative process, 158

Ritty, James J., 125
Robbins, Bud, advertising grand
    pianos, 150–151
Roentgen, Wilhelm, 87
Rogers, Will, taking action, 173
Rombauer, Irma, *The Joy of
    Cooking*, 179
Rooney, Andy, meeting deadlines,
    166
Rosseau, Jean-Jacques, word
    combinations, 117
roulette, 86
routine, changing daily, 71–74
rubber, vulcanized, 86
Rubinstein, Moshe F., scientific
    problem solving, 8–9
rules, breaking, 30, 119
Runyon, Damon, betting on a race,
    18
Russell, Bertrand
    thinking, 101
    working underground, 167
Russell, James, 64
rut, changing routine, 71–74
Ruth, Kent, new thoughts, 35

Saarinen, Eero, 120
safety pin, 115
Sagan, Carl
    children's questions, 30
    unconscious problem solution,
    170
Saint, Thomas, 181
salad, Caesar, 115
sales, improving cold sales call
    results, 142
Salk, Jonas, asking the right
    questions, 133
Sartre, Jean-Paul
    courage, 85
    self-concept, 46
    word combinations, 117
savings, increasing, 141

Schick Electric Company, 97–98
school, effect on children, 30
Schultz, Charles
    knowing the answer, 1
    running away from problems, 131
Scotch tape, 86
Scotchgard, 87
Scrabble, 64
search engines, research with
    Internet, 153
security, endangering our, 85
seeds, fun at work, 20–22
seeing, vs. looking, 75–76
self-image, performance, 43–48
self-publishing, 179
sewing machines, 180–181
shaving cream, heated, 97–98
Shaw, George Bernard, wasted
    youth, 25
Sherman, Patsy, 87
Shula, Don, coaching prowess, 117
Silverman, Hal, 162
    working with, 78, 95
simplicity, ideas, 3–4
Sinatra, Frank, word combinations,
    117
Sister Carrie, 65
smallpox vaccine, 133
Smokey Bear, 32, 39–40
SOAPRIZE, 95–96
solutions
    changing the problem definition,
        134–136
    childlike, 33
    conditioning, 10
    finding, 49, 61
    limited by assumptions, 109–110,
        113
    multiple, 41, 111–113
    problems, 38, 40, 105
    teamwork, 98–99
    verification of, 9
    visualizing, 105

See also problems
speaking, public, 44
Speculators, 37
Spencer, Percy LeBaron, 87
spirit, team, 16
spring break, student behavior,
    136–137
springboard, need for a, 152
St. Augustine, intelligo, 5
start, waiting to, 176
statement, mission, 132
steel, crucible, 125
Steffens, Lincoln, improving the
    existing world, 39
Stevens, Risë, moving the Met to
    Lincoln Center, 151
stories
    failure to implement a discovery,
        180
    writing short, 88
strategy, creative, 132
Stravinsky, Igor, 120
stubbornness, refusing to accept
    failure, 64–65
students, Florida spring break
    behavior, 136–137
subconscious, incubation in the,
    8–9
success
    failure, 61
    self-image, 45
Suits, C.G., General Electric
    research laboratories, 167
surfing, the Internet, 73,
    152–153
surrealism, 18
Svevo, Italo, forgetting, 165
Swiss Army knife, 40–41
Syrus, Publilius, forgetting who we
    are, 165

tables, radical new designs, 120
tapes, altering your self image, 48

Index

tardiness, employee, 106–107
teamwork
    friends, 94
    placing blame, 93
    team size, 99
Teflon, 87
telephone, 86
telescope, 87
    Hubble, 126
tenacity, legendary, 163
Tesla, Nikola, 163
The Act of Creation (Kostler), 7
The Babinski Reflex (Goldberg), 168
The Courage to Create (May), 114
The Dancing Wu Li Masters (Zukav), 26–27
The Grace of Great Things (Gruden), 85
The Joy of Cooking (Rombauer), 179
The Society of Mind (Minsky), 2–3
Theorem, Pierre de Fermat's Last, 163
theories
    gravity, 18
    natural selection, 19, 125
    relativity, 28
Thimonnier, Barthelemy, 181
thinking
    in images, 102–105
    lateral, 105–109
    milking cows, 167
thirteen, half of, 160–162
Thomas, Dylan, childhood perspectives, 28–29
thought, joining matricies of, 18
Thurber, James
    knowing the questions, 131
    thinking, 101
time, getting ideas, 49
Tomlin, Lily
    answering questions, 1
    individual responsibility, 93
    the rat race, 51

toys, Slinky, 87
training, employee retention, 107–109
trees, bark, 73
tunnel, light at tne end of the, 93
Twain, Mark
    answering questions, 1
    Huckleberry Finn, 179

underground, working, 167, 170
Upton, F.R., Edison's work habits, 177

vaccine, smallpox, 133
Valéry, Paul, serious people, 17
van Gogh, Vincent, breaking rules, 119
Vidal, Gore
    human problems, 131
    success and failure, 59
videos, altering your self image, 48
Virgil, self-image, 45
visualization, methods, 52–54, 56
von Helmholtz, Hermann
    getting ideas, 167
    three steps to new thoughts, 8
von Oech, Roger
    A Whack on the Side of the Head, 124
    father of invention, 17

Wakefield, Charles S., *Predator of the Universe: The Human Mind*, 9
Walker, John, 87
water, carbonated, 86
Weakley, Jeff, magazine ad resume, 69–71
Wegener, Alfred, relationships of images, 103
Westinghouse, George, 125
What Color Is Your Parachute? (Bolles), 179
"What If?", playing, 121–123

white horse game, 75
Whitman, Walt, *Leaves of Grass*, 179
Wiener, Norbert, scientific problem
    solving, 42
Wilde, Oscar, seriousness, 15
Wiles, Andrew, 163
Winner barbed wire, 177
Wise, Ernie and Eric Morecambe,
    forgetting, 165
wonder, childlike sense of, 30
wood block, finding uses for a
    2x2x2, 159
Wood, Grant, getting good ideas,
    167
work
    the assembly line, 133
    attire at, 20
    fun at, 16–22
    taking breaks, 168–169
work plan, creative, 132
workers, compensation of, 120
Wozniak, Steve, 179
Wright, Frank Lloyd, imagining
    concepts, 102

Wright, Orville and Wilbur, 88
writer's block, 29
writing
    new styles of, 120
    in teams, 94–95

x-rays, 86–87
Xerox machine, 64

Yahoo, researching on the Internet,
    153
yogurt, billboard for Knudsen, 96
Young, James Webb
    A Technique for Producing Ideas, 8
    combining old elements, 4, 7, 69
    producing ideas, 11
    specific knowledge, 152

Zen, story of Nan-in, 27
*Zen and the Art of Motorcycle
    Maintenance* (Pirzig), 29, 65
Zukav, Gary, *The Dancing Wu Li
    Masters*, 26–27

About the Author

Jack Foster was eighteen years old and working in an insurance company with about 150 other people when he got the idea to raffle off his weekly paycheck. Fifty cents a chance to win $27.50.

The first week he made a profit of $6.

The next week he had collected $53 dollars for the raffle when his boss found out what he was doing. He ordered Jack to return the money.

Then he fired him.

Ever since, Jack's been trying to come up with ideas that wouldn't get him fired.

Mostly he's succeeded.

He lucked into the advertising business fifty years ago as a writer and has been coming up with ideas ever since. Ideas for scores of companies including Carnation, Mazda, Sunkist, Mattel, ARCO, First Interstate Bank, Albertson's, Ore-Ida, Suzuki, Denny's, Universal Studios, Northrup, Hughes, Disney, Rand McNally, and Smokey Bear.

During the fifteen years Jack spent as the executive creative director of Foote, Cone & Belding in Los Angeles, it grew to be the largest advertising agency on the West Coast. He also has won dozens of advertising awards, including being named "Creative Person of the Year" by the Los Angeles Creative Club.

For seven years he helped teach an advanced advertising class at the University of Southern California that was sponsored by the American Association of Advertising Agencies, and for three years he helped teach an extension class at the University of California at Los Angeles on creating advertising.

Jack married Nancy ("The best idea," he says, "I ever had.") forty-nine years ago. They live in Santa Barbara.

About the Illustrator

 I was born in London, England. It was raining.

After fifteen years of studying Latin I decided to go into advertising.

My first job was as an apprentice at an advertising agency called Graham and Gilles. I changed the water pots for the art directors (they painted layouts with watercolours in those days) and made them tea. This was before magic markers. This was even before rubber cement—I'm that old.

It was raining. It was always raining, and I was watching my favourite programme at the time— *77 Sunset Strip*. I said, "Ah, sun, palm trees, women." My dad gave me a one-way ticket.

I met Jack Foster forty years ago at the Erwin Wasey advertising agency in Los Angeles and then again at Foote, Cone & Belding.

We worked together for about seventeen years. We had a hell of a good time.

And we had a hell of a good time doing this book.

About Berrett-Koehler Publishers

Berrett-Koehler is an independent publisher dedicated to an ambitious mission: Creating a World that Works for All.

We believe that to truly create a better world, action is needed at all levels—individual, organizational, and societal. At the individual level, our publications help people align their lives with their values and with their aspirations for a better world. At the organizational level, our publications promote progressive leadership and management practices, socially responsible approaches to business, and humane and effective organizations. At the societal level, our publications advance social and economic justice, shared prosperity, sustainability, and new solutions to national and global issues.

A major theme of our publications is "Opening Up New Space." They challenge conventional thinking, introduce new ideas, and foster positive change. Their common quest is changing the underlying beliefs, mindsets, and structures that keep generating the same cycles of problems, no matter who our leaders are or what improvement programs we adopt.

We strive to practice what we preach—to operate our publishing company in line with the ideas in our books. At the core of our approach is stewardship, which we define as a deep sense of responsibility to administer the company for the benefit of all of our "stakeholder" groups: authors, customers, employees, investors, service providers, and the communities and environment around us.

We are grateful to the thousands of readers, authors, and other friends of the company who consider themselves to be part of the "BK Community." We hope that you, too, will join us in our mission.

A BK Life Book

This book is part of our BK Life series. BK Life books change people's lives. They help individuals improve their lives in ways that are beneficial for the families, organizations, communities, nations, and world in which they live and work. To find out more, visit www.bk-life.com.

Be Connected

Visit Our Website

Go to www.bkconnection.com to read exclusive previews and excerpts of new books, find detailed information on all Berrett-Koehler titles and authors, browse subject-area libraries of books, and get special discounts.

Subscribe to Our Free E-Newsletter

Be the first to hear about new publications, special discount offers, exclusive articles, news about bestsellers, and more! Get on the list for our free e-newsletter by going to www.bkconnection.com.

Participate in the Discussion

To see what others are saying about our books and post your own thoughts, check out our blogs at www.bkblogs.com.

Get Quantity Discounts

Berrett-Koehler books are available at quantity discounts for orders of ten or more copies. Please call us toll-free at (800) 929-2929 or email us at bkp.orders@aidcvt.com.

Host a Reading Group

For tips on how to form and carry on a book reading group in your workplace or community, see our website at www.bkconnection.com.

Join the BK Community

Thousands of readers of our books have become part of the "BK Community" by participating in events featuring our authors, reviewing draft manuscripts of forthcoming books, spreading the word about their favorite books, and supporting our publishing program in other ways. If you would like to join the BK Community, please contact us at bkcommunity@bkpub.com.